T0392568

VOLUME THIRTY EIGHT

ADVANCES IN BIOMEMBRANES AND LIPID SELF-ASSEMBLY

EDITORIAL BOARD

VOLUME THIRTY EIGHT

ADVANCES IN BIOMEMBRANES AND LIPID SELF-ASSEMBLY

Edited by

ALEŠ IGLIČ
University of Ljubljana, Slovenia

MICHAEL RAPPOLT
University of Leeds, United Kingdom

PATRICIA LOSADA-PÉREZ
Université libre de Bruxelles, Belgium

Academic Press is an imprint of Elsevier
50 Hampshire Street, 5th Floor, Cambridge, MA 02139, United States
525 B Street, Suite 1650, San Diego, CA 92101, United States
The Boulevard, Langford Lane, Kidlington, Oxford OX5 1GB, United Kingdom
125 London Wall, London, EC2Y 5AS, United Kingdom

First edition 2023

ISBN: 978-0-323-99246-6
ISSN: 2451-9634

For information on all Academic Press publications
visit our website at https://www.elsevier.com/books-and-journals

Publisher: Zoe Kruze
Acquisitions Editor: Jason Mitchell
Editorial Project Manager: Palash Sharma
Production Project Manager: Abdulla Sait
Cover Designer: Miles Hitchen

Typeset by MPS Limited, India

Contents

Contributors

Maxence Berry
University of Ljubljana, Faculty of Health Sciences, Laboratory of Clinical Biophysics, Ljubljana, Slovenia; University of Poitiers, College of Fundamental and Applied Science, Poitiers, France

Matej Daniel
Department of Mechanics, Biomechanics, and Mechatronics, Faculty of Mechanical Engineering, Czech Technical University in Prague, Technická, Prague

Damjana Drobne
University of Ljubljana, Biotechnical Faculty, Nanobiology Group, Ljubljana, Slovenia

Yuri Gerelli
Institute for Complex Systems, National Research Council; Department of Physics, Sapienza University of Rome, Piazzale Aldo Moro, Rome, Italy

Zala Jan
University of Ljubljana, Faculty of Health Sciences, Laboratory of Clinical Biophysics, Ljubljana, Slovenia

Monika Jenko
MD-RI Institute for Materials Research in Medicine; Institute of Metals and Technology, Ljubljana, Slovenia

Veronika Kralj-Iglič
University of Ljubljana, Faculty of Health Sciences, Laboratory of Clinical Biophysics, Ljubljana, Slovenia

Katarína Mendová
Department of Mechanics, Biomechanics, and Mechatronics, Faculty of Mechanical Engineering, Czech Technical University in Prague, Technická, Prague

Jitka Řezníčková
Department of Mechanics, Biomechanics, and Mechatronics, Faculty of Mechanical Engineering, Czech Technical University in Prague, Technická, Prague

Santiago Daniel Salas
Instituto de Investigaciones en Fisicoquímica de Córdoba (INFIQC-UNC-CONICET), Departamento de Química Orgánica. Facultad de Ciencias Químicas, Universidad Nacional de Córdoba. Haya de la Torre y Medina Allende, Ciudad Universitaria, Córdoba, Argentina

Raquel Viviana Vico
Instituto de Investigaciones en Fisicoquímica de Córdoba (INFIQC-UNC-CONICET), Departamento de Química Orgánica. Facultad de Ciencias Químicas, Universidad Nacional de Córdoba. Haya de la Torre y Medina Allende, Ciudad Universitaria, Córdoba, Argentina

Martín Eduardo Villanueva
Experimental Soft Matter and Thermal Physics (EST) group, Department of Physics, Université libre de Bruxelles, Boulevard du Triomphe CP223, Brussels, Belgium

Preface

This volume is devoted to a very relevant and actual topic in membrane biophysics, namely, the interaction of nanoparticles with lipid membranes. The chapters contributing to this volume cover the theoretical understanding of nano-biomembrane interfaces and high-resolution nanoscale techniques to probe the extent of nanoparticle impact into bilayer structure. The lipid-based systems under study range from artificial monolayers and supported lipid bilayers to whole cell. Volume 38 consists of five chapters and is organized as follows.

Chapter 1 describes the potential effects of inorganic debris at the cellular level in patients with orthopedic implants using human umbilical vein endothelial cells (HUVEC) as a cell model. The authors evaluated the cytotoxicity of Al_2O_3 particles and ZrO_2/SiO_2 composites and their impact on inflammatory response and oxidative stress of HUVEC. The results of this study indicate increase in concentration of interleukins (IL-1β and IL-6) when cells were treated with Al_2O_3 nanoparticles. Selected inorganic nanoparticles also induce oxidative stress response, shown with increase in cholinesterase and glutathione S-transferase activity as well as lipid droplet production in in vitro HUVEC cells. The results highlight that potential negative health effects in vivo can be mitigated in the future with appropriate therapy, including antioxidants against oxidative stress induced by inorganic debris.

Chapter 2 deals with the mutual influence between topological defects and nanoparticle assembly in liquid crystals. In particular, it describes the impact of nanoparticles into defect mobility, stability, and induction of new defects. It also discusses on how nanoparticles change the stability of liquid crystal phases, as well as their thermal and optical and electro-optical properties. The chapter highlights the analogies between liquid crystal phases and biological membranes and how the study of the former can be useful to assess the interaction of nanoparticles with the latter.

Chapter 3 is a comprehensive review on the use of neutron and X-ray reflectometry to gain structural insights into solid-supported lipid bilayers and the spatial distribution of soft and hard nanoparticles either onto or within the bilayers. Several case studies are presented, highlighting the use of NR and XRR techniques to discuss the influence of nanoparticle surface charge, shape, and composition on the mode of interaction, membrane integrity, and lipid organization. The role of membrane complexity, including the presence of natural lipid and protein components, in shaping nanoparticle–membrane interactions is emphasized.

Chapter 4 presents Langmuir monolayers (LMs) as useful models for evaluating the interaction between different types of biologically relevant nanoparticles and the outer leaflet of the cell membrane. The authors provide a detailed description of the experimental setup and methodology as well as the advantages and limitations to work with LMs, in particular to perform and validate in vivo studies.

In Chapter 5, the authors provide a refined model for the preparation of platelet and extracellular vesicle-rich plasma by centrifugation of blood with maximal recovery of platelets that considers the effect of hematocrit. The chapter compares optimal plasma length obtained from the model and measurements as well.

We would like to thank all authors who contributed to Volume 38. We would also like to express our gratitude to Jason Mitchell from Elsevier in Oxford to Palash Sharma from Elsevier in New Delhi, a content team manager, and Akanksha Marwah from Elsevier in New Delhi.

Aleš Iglič
Michael Rappolt
Patricia Losada-Pérez

CHAPTER ONE

Interaction of inorganic debris particles with cells

Zala Jan[a,*], Damjana Drobne[b], Monika Jenko[c,d], and Veronika Kralj-Iglič[a]

[a]University of Ljubljana, Faculty of Health Sciences, Laboratory of Clinical Biophysics, Ljubljana, Slovenia
[b]University of Ljubljana, Biotechnical Faculty, Nanobiology Group, Ljubljana, Slovenia
[c]MD-RI Institute for Materials Research in Medicine, Ljubljana, Slovenia
[d]Institute of Metals and Technology, Ljubljana, Slovenia
*Corresponding author. e-mail address: zala.jan@zf.uni-lj.si

Contents

Abstract

After human umbilical vein endothelial cells (HUVEC) were treated with milled particles simulating debris involved in sandblasting of orthopaedic implants (OI) for 24 h, inflammatory and oxidative stress-related parameters as well as cytotoxic response were monitored. Three different types of particles were used: abrasives [corundum—(Al_2O_3), used corundum retrieved from removed OI (u. Al_2O_3), and zirconia/silica composite (ZrO_2/SiO_2)]. Scanning electron microscopy (SEM) was used for cell morphological changes observation. Inflammatory process was monitored by measuring concentration of Interleukin(IL)-6, IL-1β and Tumour Necrosis Factor(TNF)-α with enzyme-linked immunosorbent assay (ELISA). Oxidative stress was monitored by spectrophotometry assessing activity of cholinesterase (ChE) and glutathione *S*-transferase (GST) and assessing reactive oxygen species (ROS) as well as lipid droplets (LD) by flow cytometry (FCM). FCM was used also for apoptosis assessment. For treated and untreated cells, the extent of cell detachment from glass discs and the budding of the cell membrane were similar. In cells treated with particles, the concentrations of IL-1β and of IL-6 were higher, indicating inflammatory response of the treated cells. ChE activity increased after cell treatment with u. Al_2O_3 and ZrO_2/SiO_2. Increased GST activity was found after cells were treated with ZrO_2/SiO_2. After cell

ISSN 2451-9634, https://doi.org/10.1016/bs.abl.2023.08.001

treatment with u. Al_2O_3, LD quantity increased and ROS quantity in treated cells was comparable with the one in untreated cells. Cytotoxic effect of the particles was not detected when cells were treated with u. Al_2O_3. Bioactivity of the tested materials in concentrations added to in vitro cell culture indicates a response of the human body to OI.

1. Introduction

Biocompatibility and relevance of materials used for orthopaedic implants (OI) is becoming more and more important due to increased life expectancy. OI materials must be biologically acceptable not to cause unwanted local tissue reactions and robust enough to allow patients to perform daily life activities [1]. Materials recently used for joint replacement are well adapted in the body if they are in the bulk form, mechanically stable and sterile [1]. Yet, metallic wear particles released into the tissue that surrounds OI can result in formation of fibrotic tissue around surgical implants [2]. With damage of the tissue, inflammatory processes can be triggered and inflammatory processes can cause fibrosis through different pathways [3]. The choice of the materials used for OI and surface amplification are key to the longevity of the endoprosthesis. Reduction of revision surgeries lower risk to patients, also regarding thromboembolic events, infection, dislocation, and death [4]. Common reason for revision surgeries is aseptic loosening of the OI, which is the result of excessive wear of OI that produces particle debris (and consequently, osteolysis). Aseptic loosening of the OI causes pain and reduced mobility [5,6].

Metal, polyethylene and ceramic are the most commonly used materials for OI. Different materials and their combinations have different survival rates and patients report different problems (reviewed in [7]). A 6-year (mid-term) follow-up study including 310 hips with ceramic head and liner prostheses showed that 99.0% of the hips had not been associated with re-operation and there was no radiological evidence of osteolysis or loosening [8]. In this study, ceramics showed promising results regarding wear and related loosening. It was reported that alumina–alumina ceramic OI generated 400 times fewer wear particles than metal-polyethylene OI, which resulted in a lower rate of periprosthetic osteolysis in alumina–alumina OI [9]. Zirconia-toughened alumina ceramic was found to release 1 μg/year of wear debris into circulation and surrounding tissue, which is considered very low [10].

Microscopic properties of the interfaces between the prosthesis parts as well as between the prosthesis and tissues are the important aspects when talking about improving the quality of the prostheses. Most of the debris in ceramics consist of particles sized between 0.1 and 10 μm, yet also larger particles up to 1 mm were observed [11]. Along with the toxic effects, inflammatory and oxidative stress responses of the cells are important. It was summarised in the review on the response of cells that, upon biomaterial implantation, a sequence of events is initiated with an injury, followed by blood–material interactions, provisional matrix formation and acute innate inflammatory response acting on monocytes, fibroblasts, osteoblasts, osteoclasts, and mesenchymal stem cells [12]. The activation of macrophages was suggested as the dominant mechanism in periprosthetic inflammation [13]. It begins with interaction of the particles on membrane receptors and is followed by the release of pro-inflammatory cytokines [e.g., tumour necrosis factor (TNF)-α, IL-1β, IL-6], growth factors (macrophage colony stimulating factor 1—M-CSF), pro-osteoclastic factors (receptor activator of nuclear factor kappa B ligand—RANKL) and chemokines (e.g., IL-8, macrophage inflammatory protein—MIP-1α, monocyte chemoattractant protein—MCP-1). Moreover, phagocytosis of wear debris takes place [12]. It was suggested that inefficient phagocytosis with excessive production of inflammatory mediators may lead to sustained inflammation and, eventually, fibrotic changes [3,14].

Oxidative stress-response of the cells depends on biocompatibility of the materials used for OI. Along with signalising, ROS are important in the debris degradation process [11]. Since cells and particles produce oxidants, it is believed that oxidative stress represents most direct communication between cells and OI debris. Continuing oxidative stress defence of immune cells can lead to long-term exposure of cells to oxidants and therefore to chronical inflammation. As a result, biocompatibility and function of OI is lost [15]. Oxidative stress can also accelerate degradation of OI material which results in higher amount of debris released in tissue surrounding the OI. Macrophages have key role in debris disposal in during this process also oxidants are formed [11]. We can conclude that debris directly and indirectly impacts oxidants formation. It was reported that ceramic OI debris significantly increase lipid peroxidation and lower antioxidant enzymes in tissue surrounding the OI [16].

Debris, released from OI can be toxic for human cells in two ways: chemical, when soluble ions and monomers are released into the surrounding tissue and blood circulation and/or mechanical insoluble particles are recognised as foreign objects [17]. Cytotoxicity is lower for insoluble particles [17], among which are Al_2O_3 and ZrO_2/SiO_2, used in our research.

With the aim of better understanding the mechanisms underlying the effects of the sandblasting debris contamination on osteointegration, it is important to study the effect of debris on cells. In this study we address the effect of three types of particles: Al_2O_3—white alumina; u. Al_2O_3—white alumina previously used in the process of sandblasting of OI; and ZrO_2/SiO_2—zirconia/silica composite cytotoxicity and their impact on inflammatory response and oxidative stress of HUVEC.

2. Methods

HUVEC cells were treated with three different types of particles: Al_2O_3, white alumina; u. Al_2O_3, white alumina previously used in the process of sandblasting of OI; and ZrO_2/SiO_2 in three different concentrations: 10, 50 and 100 μg/mL. Particles were obtained from FerroECOBlast, Dolenjske Toplice, Slovenija. Particles were milled and analysed with measuring of Zeta potential and Dynamic Light Scattering (DLS), also, abrasives were characterised. After treatment, inflammation and oxidative stress markers were measured and cell apoptosis was analysed. Scheme of methods (Fig. 1) is added, additional information can be found in Jan et al. (2023) [18].

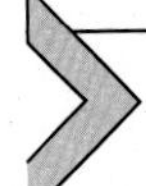

3. Results

3.1 Inflammatory response of the treated HUVEC cells

To assess the inflammatory response of HUVEC, IL-6, IL-1β and TNF-α were measured in conditioned media of HUVEC after 24 h exposure of the cells to three different types of particles at three concentrations (10, 50 and 100 μg/mL). For positive control, 10 μM lipopolysaccharide (LPS) was used. There seem to be an increase in IL-6 with respect to the untreated control cells was observed in samples treated with u. Al_2O_3 and ZrO_2/SiO_2 at all three concentrations (Fig. 2A). Only in case of treatment with u. Al_2O_3, increase in IL-6 concentration was statistically important. In samples treated with unused Al_2O_3 the effect was the least; for the lowest concentration the effect

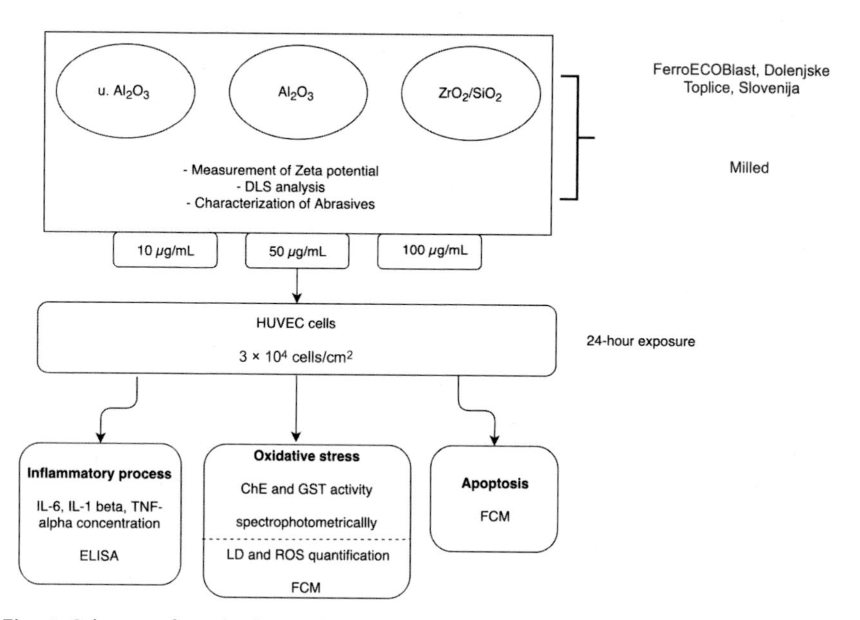

Fig. 1 Scheme of methods used in the study. DLS: dynamic light scattering, HUVEC: human umbilical vein endothelial cells, IL: interleukin, ChE: cholinesterase, GST: glutathione S-transferase, LD: lipid droplets, ROS: reactive oxygen species, FCM: flow cytometry.

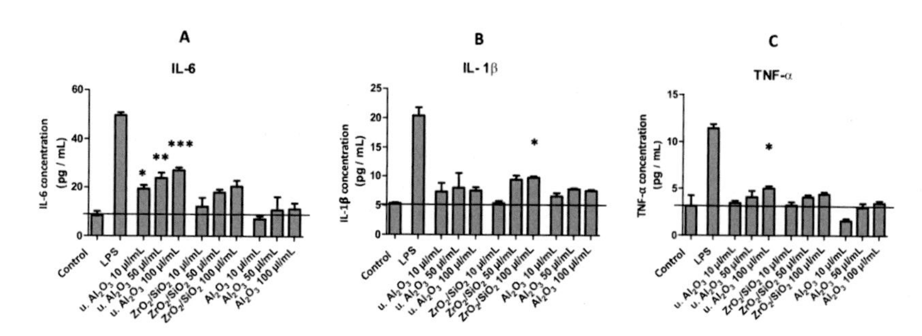

Fig. 2 Concentration of IL-6 (A), IL-1β (B) and TNF-α (C) in conditioned media of cells treated with three different types of particles (u. Al_2O_3, ZrO_2/SiO_2 and Al_2O_3) and 10 μM LPS—positive control. Particles were administered at three concentrations (10 μg/mL, 50 μg/mL and 100 μg/mL). The horizontal line denotes the average concentration of negative control samples (untreated cells). Experiments were performed in duplicate, and bars represent the standard deviations. Asterisks denote statistically significant differences with respect to the control. IL: interleukin, TNF-α: tumour necrosis factor α, LPS: lipopolysaccharide, $^{*}p < 0.05$, $^{**}p < 0.01$ and $^{***}p < 0.001$. *Adapted from [18].*

was within the experimental error of the control, while the higher concentrations of Al_2O_3 likewise increased the concentration of IL-6 in the conditioned medium (Fig. 2A). These results indicate that all types of particles tested induced an increase of IL-6 concentration in the conditioned media. The effect of the particles on the concentration of IL-1β in the conditioned media was less pronounced than that of IL-6; it stayed within the experimental error for u. Al_2O_3 particles with concentrations 10 and 50 μg/mL (Fig. 2B). However, at higher concentrations of ZrO_2/SiO_2 and unused Al_2O_3, an increase in the concentration of IL-1β was noted (Fig. 2B). TNF-α concentration increased in samples treated with ZrO_2/SiO_2 at 50 μg/mL and 100 μg/mL and in samples treated with u. Al_2O_3 at all three concentrations (Fig. 2C), but increase was not statistically important. For both ZrO_2/SiO_2 and u. Al_2O_3, a concentration-dependent trend was observed. TNF-α concentration was not increased in samples treated with unused Al_2O_3 (Fig. 2C). As expected, treatment with LPS resulted in evident increase in cytokine concentration.

4. Oxidative stress of the treated HUVEC cells

To assess the oxidative stress response of HUVEC, activities of ChE and GST as well as quantities of ROS and LD were measured after 24 h exposure of HUVEC to three different types of particles at three concentrations (10 μg/mL, 50 μg/mL and 100 μg/mL). ChE and GST activities were expressed as activity in nmol/min/ng/proteins. ROS and LD production were expressed by the fold change of median fluorescence intensities of the respective dyes in comparison to control cells. The average ChE activities of all tested samples, except 50 μg/mL and 100 μg/mL of unused Al_2O_3, were higher than the average ChE activity of the control; however, the experimental errors were rather large. The only statistically significant difference with respect to the control was observed with 100 μg/mL of u. Al_2O_3 (Fig. 3A). The activity of GST was higher in samples treated with u. Al_2O_3 in a concentration-dependent way (Fig. 3B). A slight increase of GST activity was observed when cells were treated with 100 μg/mL of ZrO_2/SiO_2 and with u. Al_2O_3 at higher concentrations (Fig. 3B). The amount of ROS was increased in all samples except for two (samples treated with u. Al_2O_3 at 10 μg/mL and with ZrO_2/SiO_2 at 50 μg/mL) (Fig. 3C). For u. Al_2O_3, a concentration-dependent trend was observed (Fig. 3C). The number of LDs

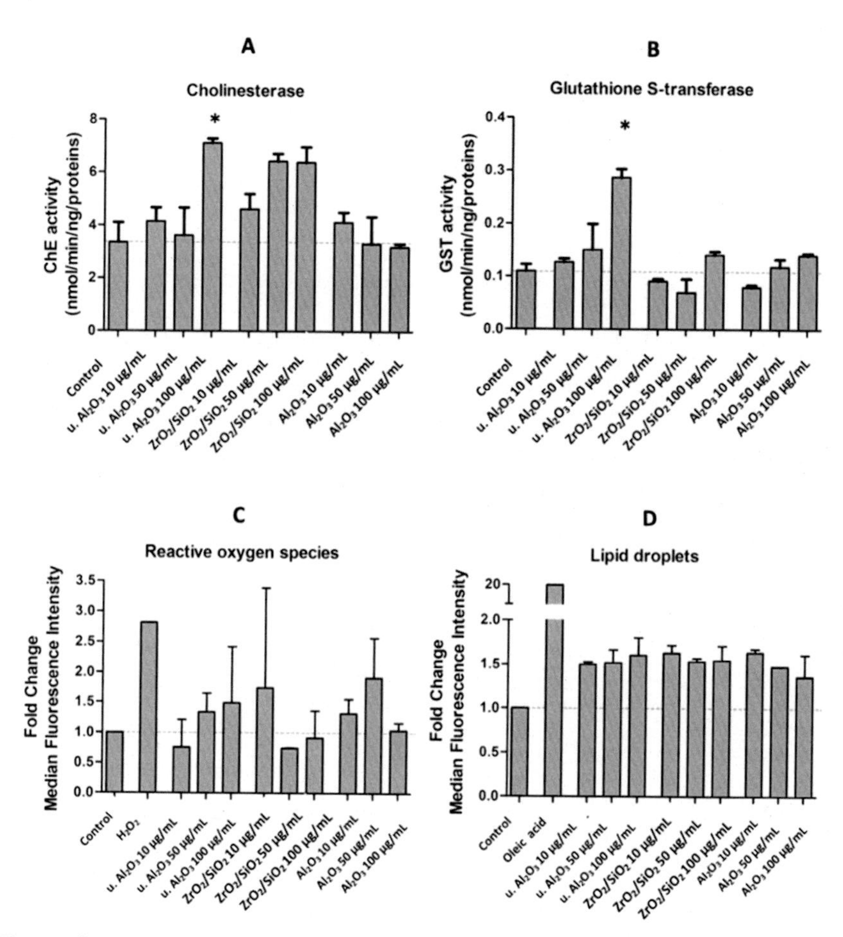

Fig. 3 ChE activity (A), GST activity (B), fold change of median fluorescence intensity of ROS (CM-H_2DCFA) (C) and LD (BODIPY 483/503) (D) in comparison to controls of HUVEC cells treated with either of the three different types of particles (u. Al_2O_3, ZrO_2/SiO_2 and Al_2O_3) at three concentrations (10 µg/mL, 50 µg/mL and 100 µg/mL), or the positive controls (5 mM H_2O_2 and 75 µM oleic acid). Experiments were performed in triplicate (A,B) or duplicate (C,D) and bars represent the standard deviations. Asterisks denote statistically significant differences with respect to the control, $*p < 0.05$. Fig. 3 was already published in [18].

was increased in all treated samples in comparison to untreated samples (Fig. 3D). No clear LD concentration-dependent trend was observed (Fig. 3D). A clear increase in ROS and LD production was successfully induced in positive control samples after treatment with 5 mM H_2O_2 and 75 µM oleic acid.

5. Cytotoxicity

We assessed apoptosis of the cells treated with u. Al_2O_3 particles since u. Al_2O_3 seem to be the most bioactive in terms of ChE, GST and IL-6 production. Apoptosis may play an important role in the regulation of inflammation or be the result of inflammation in cells. Fig. 4 shows both FSC/SSC dot plots and Annexin V fluorescence histograms (marker for apoptosis) of untreated cells (negative control), Staurosporine-treated cells (positive control) and for cells treated with u. Al_2O_3 in three different concentrations (10, 50 and 100 μg/mL). Staurosporine effectively expanded the apoptotic population during the 24 h treatment compared to untreated control cell apoptotic population (69% vs 5%). In contrast, no significant differences are visible between the negative control and particle-treated

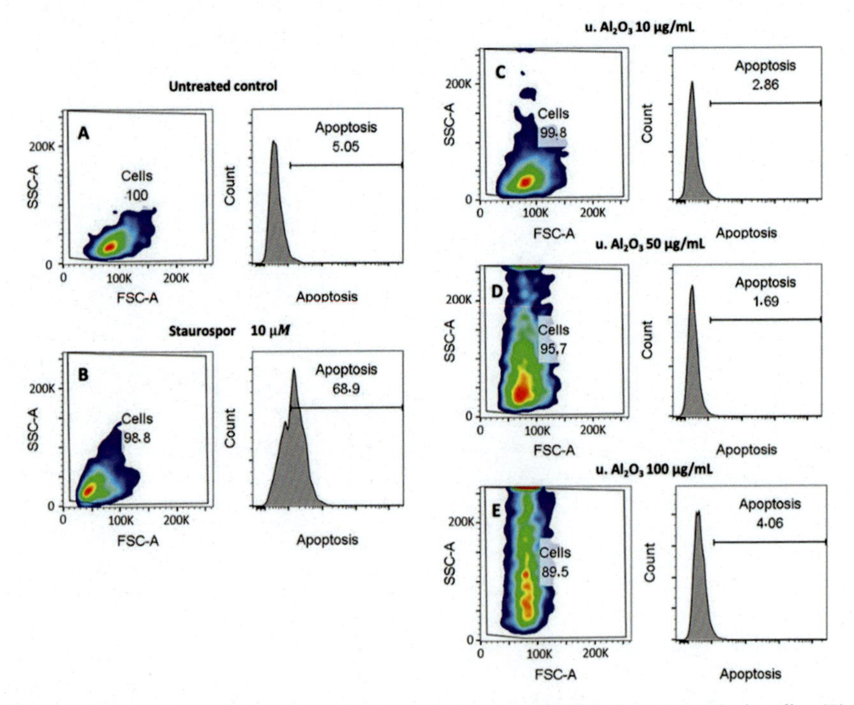

Fig. 4 Cytotoxicity effect of u. Al_2O_3 particles on HUVEC. (A) untreated cells, (B) positive control for apoptosis (cells treated with 10 μM Staurosporine), (C) HUVEC treated with 10 μg/mL of u. Al_2O_3, (D) HUVEC treated with 50 μg/mL of u. Al_2O_3, (E) HUVEC treated with 100 μg/mL of u. Al_2O_3. SSC-A—side scatter measurements; information on internal complexity (granularity), FSC-A—forward scatter; information on cell size. The numbers in boxes represent percent of cells in the respective gates (left) and percent of apoptotic cells (right). Fig. 4 was already published in [18].

samples (Panels C–E), indicating that no apoptosis was induced upon 24 h of incubation with u. Al_2O_3. In the dot plots shown in Fig. 4, a dose-dependent increase of the SSC signal can be observed. Increase in cell granularity could be a consequence of particle endocytosis or LD/intracellular vesicle formation. This would be in line with the data presented in Fig. 2, which show an increase in inflammatory response of the cells.

6. Discussion

The purpose of this research was to monitor impact of inorganic (nano)particles used for OI sandblasting (u. Al_2O_3, ZrO_2/SiO_2 in Al_2O_3) in three different concentrations (10, 50 in 100 μg/mL) to HUVEC, exposed in vitro for 24 h. We assessed impact of (nano)particles on inflammatory process, and oxidative stress, also impact of the particles on apoptotic was of interest.

After OI is implanted in the human body osseointegration, bone ingrowth into an OI is important [19]. Osseointegration is accelerated by adding different composites to the OI and sandblasting of the OI. For start and effectiveness of osseointegration and therefore functionality of the OI, the bone in contact with OI must remain homoeostasis. Bone homeostasis is controlled with communication between bone cells and communication of bone cells with OI material. If the process is successful, growth and differentiation of bone cells is stimulated which results in ingrowth of bone into OI [20].

When monitoring inflammatory process of treated cells, particle concentration dependent increase in IL-6 concentration was observed. There was a statistically important increase of IL-6 concentration when cells were exposed to all concentrations of u. Al_2O_3. Increase of IL-1β concentration was statistically important when cells were treated with 100 μg/mL ZrO_2/SiO_2 (Fig. 2). Statistically important increase was observed also for TNF-α concentration when cells were treated with 100 μg/mL u. Al_2O_3. Treatment of the cells with LPS provoked increase of cytokine concentration which was approximately 2-fold higher than in cells treated with any (nano)particles in any concentration. In 1980 Evans in Thomas [2] reported about common occurrence of fibrosis tissue surrounding OI. Later, researchers connected occurrence of fibrosis tissue with migration of the metal particles from OI into the surrounding healthy and fibrosis tissue [21,22]. Yet little is known about impact of ceramic (nano)particles on inflammatory process in human body. In their research, Malem et al. [23] monitored a patient, after implantation of ceramic OI. Patient reported about pain and audible noise five years after hip replacement. After the revision surgery OI was observed for one year and it was found that higher concentrations of debris are being released from OI—daily 10 mm^3. Campbell et al. [23] reported about presence of pseudotumors six years after implantation of ceramic OI. Jamieson et al. [23] reported about THP-1 macrophages in vitro exposed to oxide ceramics nanoparticles. Nanoparticles were removed by phagocytosis, which impacted inflammation process of the cells. They also reported about significant increase in cell IL-1β production. As part of the in vivo study Bertrand et al. [24] reported about changes in fibrosis tissue in connection with release of ceramic OI debris released in surrounding tissue. Connection between risk for fibrosis tissue formation and period elapsed after OI implantation was found. Because of that, they assumed ceramic debris is bioactive. Reason for fibrosis tissue formation can be long-term inflammation of the peripheral blood cells and inflammatory response of fibroblasts as the response of the cells to ceramic OI debris. Sterner et al. [25] studied inflammatory response of macrophages-like cells treated with (nano)particles Al_2O_3 in ZrO_2 by monitoring changes in TNF-α concentration. They reported 4-fold higher TNF-α concentration in cells, treated with Al_2O_3 compared to concentration in untreated control cells.

In contrast to the results listed above, Warashina et al. [26] reported that OI debris did not provoke inflammatory response. They monitored cytokine concentration in patient after one week after implantation of Al_2O_3 and ZrO_2 coated OI. IL-1β, IL-6 and TNF-α concentrations were not elevated in tissue surrounding OI there was also no visible osteolysis. Since the authors of the study monitored in vivo conditions, we suppose particle concentrations were lower as particle concentrations used in our study. Bylski et al. [27] did not observe increase in TNF-α concentration in THP-1 monocytes after treatment with Al_2O_3 particles. Sterner et al. [25] reported no impact of ZrO_2 particles on TNF-α concentration in macrophages-like cells. Contradictory results of the different studies show the need for further research of the OI debris impact on inflammatory processes in human body. Also, it would be more applicable to monitor impact of (nano)particles on cells after longer period of time as the *in vivo* cells are exposed to OI debris even for a few decades.

When monitoring oxidative stress, we found statistically important increase of ChE and GST activity in cells, treated with 100 μg/mL u. Al_2O_3 (Fig. 3).

We also indicated increase in LD formation in cells, treated with any type of particles in any concentration (Fig. 3), which can indicate the response of the cells to oxidative stress. LD can help in protection of the cells from oxidative stress with reduction of the ROS, ChE activity is also connected with cell response to oxidative stress [28]. ROS formation was not statistically higher in treated cells in comparison to ROS formation in untreated cells.

Most of the researchers report impact of ingested Al_2O_3 on cell oxidative stress, overall, the in vivo and in vitro studies monitoring oxidative stress as result to OI debris are limited. Ye and Shi [29] reported about significantly increased ROS formation in 3T3-E1 mouse cells with the osteoblast's characteristics. Li et al. [30] monitored impact of Al_2O_3 nanoparticles on nematodes *Caenorhabditis elegans*, treated with nanoparticle concentrations between 0.01 and 23.1 mg/L for 10 days. They reported about higher ROS formation and lower ability of the cells to defend themselves from the ROS increase. Also, our results show increase in LD and ROS formation when cells were treated with different types of inorganic (nano)particles used for OI sandblasting. It is possible that contact between used (nano)particles and cells provoke oxidative stress and therefore LD formation to limit ROS. In contrary with above listed results and results from our study, Kim et al. [31] did not report about increased ROS formation in A549 cells even when the cells were exposed to 500 μg/mL Al_2O_3 particles for 72 h.

With our study we did not observe cell apoptosis when cells were treated with selected (nano)particles. (Nano)particles which migrate from OI in surrounding tissue and blood and can dissolve are more cytotoxic, more cytotoxic seem to be metallic particles [17]. Al_2O_3 (nano)particles do not dissolve which supposedly lower their cytotoxicity. Yamamoto et al. [17] reported that also particle shape has impact on their cytotoxicity. The higher cytotoxicity has particles with dendritic shape, followed by needle-shaped particles and the least cytotoxic are round particles. Radziun et al. [32] studied cytotoxicity of Al_2O_3 nanoparticles and concluded that used particles can penetrate membrane of L929 mouse fibroblast cells, yet viability of the cells was not lowered. Similar results were represented also by Jamieson et al. [33], who studied impact of Al_2O_3 (nano)particles on THP-1 macrophages. Even though cells ingest (nano)particles with phagocytosis, there were no cytotoxic impact. Also, Kim et al. [31] did not find cytotoxic effect of 500 μg/mL Al_2O_3 particles after 72-hour exposure. Those results are in line with our results since we also found no apoptotic effects when treating cells with selected (nano)particles for 24 h.

In contrast with our results, Catelas et al. [34] reported on higher apoptosis in J774 macrophages when they were treated with Al_2O_3 particles (with diameter of 4.5 μm) and somewhat lower apoptosis when cells were treated with smaller Al_2O_3 particles (with diameter of 0.6; 1.3 and 2.4 μm). Also, Yamamoto et al. [17] reported similar results. Olivier et al. [35] tested cytotoxic effect of Al_2O_3 particles on J774 macrophages and L929 fibroblast cells. Particles had higher cytotoxic effect on macrophages than on fibroblasts. Particles had impact on cell apoptosis and necrosis. Ye et al. [29] reported about significantly lower viability of 3T3-E1 mouse osteoblastoma like cells when treated with ZrO_2 nanoparticles. Ceramic particles can have different cytotoxic effects on different types of cells since some types of cells are more sensitive.

Cells were treated with selected (nano)particles used for sandblasting of OI for 24 h, yet debris from OI can be released in surrounding tissue and blood circulation for years even decades. Impact of debris particles on the cells depends on time of exposure, type of material used for OI and sandblasting (metallic (nano)particles seem to be less biocompatible than ceramic), shape (the least biocompatible are dendritic shaped particles, followed by needle-shaped ones and the most biocompatible are round-shaped particles), and solubility (insoluble particles less cytotoxic) of the debris particles.

7. Conclusions

Ceramic was considered to be inert material, yet more and more studies give information about capability of evoking pro-inflammatory responses of in vitro cell lines. Also, results of our study indicate increase in concentration of IL-1β and IL-6 when cells were treated with u. Al_2O_3 and Al_2O_3 (nano)particles. Selected inorganic (nano)particles also induce oxidative stress response, shown with increase in ChE and GST activity as well as lipid droplet (LD) production in in vitro HUVEC cells. Selected inorganic (nano)particles did not induce apoptosis even when treated with highest concentration—100 μg/mL. Also, there was no impact on morphological characteristics of HUVEC cells observed. Cells did not detach from the glass disk surface, and there the budding of the cell membrane was like the budding of untreated cells. Model measurements in vitro can offer data on the response of cells when in contact with ceramic orthopaedic implants debris, however, it is possible that concentrations of corundum ceramic (nano)particles in experiments are higher than in in vivo conditions. Also, in vitro conditions cannot reflect everything that is happening in human body. Therefore, studies on patients should be also performed by monitoring concentration of particles in blood and inflammation processes over time.

Funding

This research was funded by Slovenian Research Agency, grant numbers P2-0089, P2-0132, P2-0232, P3-0388, J2-3040, J2-3043, J2-3046, J2-4427, J2-4447, J3-3066, J3-3079, J7-4420, L3-2621, EUREKA IMPLANT BLASTER C2130–30–090033, National Research, Development and Innovation Office (Hungary) SNN 138407 and BI-FR/23–24-PROTEUS-005, PR-12039.

References

[1] E. Gibon, D.F. Amanatullah, F. Loi, J. Pajarinen, A. Nabeshima, Z. Yao, et al., The biological response to orthopaedic implants for joint replacement: part I: metals, J. Biomed. Mater. Res. Part. B Appl. Biomater. 105 (2017) 2162–2173.
[2] E.J. Evans, I.T. Thomas, The in vitro toxicity of cobalt-chrome-molybdenum alloy and its constituent metals, Biomaterials 7 (1986) 25–29.
[3] M. Mack, Inflammation and fibrosis, Matrix Biol. 68–69 (2018) 106–121.
[4] S. Badarudeen, A.C. Shu, K.L. Ong, D. Baykal, E. Lau, A.L. Malkani, Complications after revision total hip arthroplasty in the medicare population, J. Arthroplast. 32 (2017) 1954–1958.
[5] J.M. Mirra, R.A. Marder, H.C. Amstutz, The pathology of failed total joint arthroplasty, Clin. Orthop. Relat. Res. 170 (1982) 175–183.
[6] A. Kienzle, S. Walter, P. Von Roth, M. Fuchs, T. Winkler, M. Müller, High rates of aseptic loosening after revision total knee arthroplasty for periprosthetic joint infection, JBJS OA 5 (2020) e20.00026.
[7] D. Hannouche, M. Hamadouche, R. Nizard, P. Bizot, A. Meunier, L. Sedel, Ceramics in total hip replacement, Clin. Orthop. Relat. Res. 430 (2005) 62–71.
[8] Y.K. Lee, Y.C. Ha, J.I. Yoo, W.L. Jo, K.C. Kim, K.H. Koo, Mid-term results of the BIOLOX delta ceramic-on-ceramic total hip arthroplasty, Bone Jt. J. 99-B (2017) 741–748.
[9] P. Bizot, R. Nizard, M. Hamadouche, L. Sedel, Prevention of wear and osteolysis: alumina-on-alumina bearing, Clin. Orthop. Relat. Res. 393 (2001) 85–93.
[10] J.M. Dorlot, P. Christel, A. Meunier, Wear analysis of retrieved alumina heads and sockets of hip prostheses, J. Biomed. Mater. Res. 23 (1989) 299–310.
[11] P.A. Mouthuy, S.J.B. Snelling, S.G. Dakin, L. Milković, A. Gašparović, A.J. Carr, et al., Biocompatibility of implantable materials: an oxidative stress viewpoint, Biomaterials 109 (2016) 55–68.
[12] N.J. Hallab, J.J. Jacobs, Biologic effects of implant debris, Bull. NYU Hosp. Jt. Dis. 67 (2009) 182–188.
[13] M. Couto, D.P. Vasconcelos, D.M. Sousa, B. Sousa, F. Conceição, E. Neto, et al., The mechanisms underlying the biological response to wear debris in periprosthetic inflammation, Front. Mater. 7 (2020) 274.
[14] T.A. Wynn, K.M. Vannella, Macrophages in tissue repair, regeneration, and fibrosis, Immunity 44 (2016) 450–462.

[15] T.H. Lin, Y. Tamaki, J. Pajarinen, H.A. Waters, D.K. Woo, Z. Yao, et al., Chronic inflammation in biomaterial-induced periprosthetic osteolysis: NF-κB as a therapeutic target, Acta Biomater. 10 (1) (2014) 1–10.
[16] I. Ozmen, M. Naziroglu, R. Okutan, Comparative study of antioxidant enzymes in tissues surrounding implant in rabbits, Cell Biochem. Funct. 24 (3) (2006) 275–281.
[17] A. Yamamoto, R. Honma, M. Sumita, T. Hanawa, Cytotoxicity evaluation of ceramic particles of different sizes and shapes, J. Biomed. Mater. Res. Part A 68 (2004) 244–256.
[18] Z. Jan, M. Hočevar, V. Kononenko, S. Michelini, N. Repar, M. Caf, B. Kocjančič, et al., Inflammatory, oxidative stress and small cellular particle response in HUVEC induced by debris from endoprosthesis processing, Materials 16 (2023) 3287.
[19] R. Brånemark, P.I. Brånemark, B. Rydevik, R.R. Myers, Osseointegration in skeletal reconstruction and rehabilitation: a review, JRRD 38 (2) (2001) 175–181.
[20] H. Terheyden, N.P. Lang, S. Bierbaum, B. Stadlinger, Osseointegration-communication of cells, Clin. Oral. Implant. Res. 23 (10) (2012) 1127–1135.
[21] D. Gambera, S. Carta, E. Crainz, M. Fortina, P. Maniscalco, P. Ferrata, Metallosis due to impingement between the socket and the femoral head in a total hip prosthesis. A case report, Acta Biomed. 73 (5–6) (2002) 85–91.
[22] P. Hernigou, F. Roubineau, C. Bouthors, C.H. Flouzat-Lachaniette, What every surgeon should know about ceramic-on-ceramic bearings in young patients, EFORT Open Rev. 1 (4) (2016) 107–111.
[23] Malem D., Nagy M.T. Ghosh S. Shah B. Catastrophic failure of ceramic-on-ceramic total hip arthroplasty presenting as squeaking hip, BMJ Case Rep. 2013 (2013) bcr2013008614.
[24] J. Bertrand, D. Delfosse, V. Mai, F. Awiszus, K. Harnisch, C.H. Lohmann, Ceramic prosthesis surfaces induce an inflammatory cell response and fibrotic tissue changes, Bone Jt. J. 100-B (2018) 882–890.
[25] T. Sterner, N. Schütze, G. Saxler, F. Jakob, C.P. Rader, Effects of clinically relevant alumina ceramic particles, circonia ceramic particles and titanium particles of different sizes and concentrations on TNFα release in a human monocytic cell line, Biomed. Eng. 49 (12) (2004) 340–344.
[26] H. Warashina, S. Sakano, S. Kitamura, K.I. Yamauchi, J. Yamaguchi, N. Ishiguro, et al., Biological reaction to alumina, zirconia, titanium and polyethylene particles implanted onto murine calvaria, Biomaterials 24 (21) (2003) 3655–3661.
[27] D. Bylski, C. Wedemeyer, J. Xu, T. Sterner, F. Löer, M. Von Knoch, Alumina ceramic particles, in comparison with titanium particles, hardly affect the expression of RANK-, TNF-alpha-, and OPG-mRNA in the THP-1 human monocytic cell line, J. Biomed. Mater. Res. Part A 89 (2009) 707–716.
[28] Z. Jan, M. Drab, D. Drobne, A.B. Zavec, M. Benčina, B. Drašler, et al., Decrease in cellular nanovesicles concentration in blood of athletes more than 15 h after marathon, Int. J. Nanomed. 16 (2021) 443–456.
[29] M. Ye, B. Shi, Zirconia nanoparticles-induced toxic effects in osteoblast-like 3T3-E1 cells, Nanoscale Res. Lett. 13 (2018) 353.
[30] Y. Li, S. Yu, Q. Wu, M. Tang, Y. Pu, D. Wang, Chronic Al2O3-nanoparticle exposure causes neurotoxic effects on locomotion behaviors by inducing severe ROS production and disruption of ROS defense mechanisms in nematode Caenorhabditis elegans, J. Hazard. Mater. 219–220 (2012) 221–230.
[31] I.S. Kim, M. Baek, S.J. Choi, Comparative cytotoxicity of Al_2O_3, CeO_2, TiO_2 and ZnO nanoparticles to human lung cells, J. Nanosci. Nanotechnol. 10 (5) (2010) 3453–3458.

[32] E. Radziun, J.D. Wilczyńska, I. Książek, K. Nowak, E.L. Anuszewska, A. Kunicki, et al., Assessment of the cytotoxicity of aluminium oxide nanoparticles on selected mammalian cells, Toxicol. In Vitro 25 (2011) 1694–1700.
[33] S. Jamieson, A. Mawdesley, D. Deehan, J. Kirby, J. Holland, A. Tyson-Capper, Inflammatory responses to metal oxide ceramic nanopowders, Sci. Rep. 11 (2021) 10531.
[34] I. Catelas, A. Petit, D.J. Zukor, R. Marchand, L. Yahia, O.L. Huk, Induction of macrophage apoptosis by ceramic and polyethylene particles in vitro, Biomaterials 20 (1999) 625–630.
[35] V. Olivier, J.L. Duval, M. Hindié, P. Pouletaut, M.D. Nagel, Comparative particle-induced cytotoxicity toward macrophages and fibroblasts, Cell Biol. Toxicol. 19 (3) (2003) 145–159.

CHAPTER TWO

Interactions between biomembrane embedded nanoparticles mediated by lipid bilayer

Matej Daniel*, Jitka Řezníčková, and Katarína Mendová
Department of Mechanics, Biomechanics, and Mechatronics, Faculty of Mechanical Engineering, Czech Technical University in Prague, Technická, Prague
*Corresponding author. e-mail address: matej.daniel@gmail.com

Contents

Abstract

Biological membranes are complex and heterogeneous, consisting of various molecules in their hydrophobic interior and hydrophilic surface. While traditionally viewed as passive platforms for protein diffusion, recent research highlights the importance of nonspecific lipid properties in protein-mediated functions. Lipid-protein interactions impact all membrane functions but are challenging to study due to the flexibility and instability of membrane proteins. To overcome these challenges, artificial nanoparticles are used to investigate membrane-lipid interactions.

Nanoparticles offer advantages for membrane studies: their rigid geometry and surface chemistry allow controlled membrane deformation, and metal-containing nanoparticles enhance imaging visibility. Additionally, nanoparticles are simpler to model than complex proteins with diverse conformations.

Nanoparticles interact with membranes in different ways, including surface adsorption, translocation, and incorporation into the hydrophobic core. This chapter focuses on the embedding of hydrophobic nanoparticles within lipid bilayers,

Advances in Biomembranes and Lipid Self-Assembly, Volume 38
ISSN 2451-9634, https://doi.org/10.1016/bs.abl.2023.09.001

resembling the inclusion of transmembrane proteins. It explores the properties enabling nanoparticle embedding and examines their influence on membrane functions. Furthermore, the chapter explores how cell membranes modulate interactions among distributed nanoparticles.

1. Introduction

Biological membranes are highly heterogeneous structures with a variety of molecules existing in and across their hydrophobic interior and hydrophilic surface [1]. The standard model considers a fluid lipid bilayer as a platform for the diffusion of membrane proteins. From this perspective, lipid membranes create a passive environment where proteins can perform vital biological functions. An alternative viewpoint is that nonspecific lipid properties are directly implicated in protein-mediated functions [2]. There is ongoing evidence that lipid-protein interactions affect all membrane functions [3]. However, studying membrane-lipid interactions is challenging due to the flexibility, instability, and various conformations that membrane proteins can assume [4]. Since the interaction between lipids and proteins is primarily non-specific [5], one approach to studying these interactions is through the use of artificial nanoparticles introduced into model lipid membranes.

The utilization of nanoparticles (NPs) for studying membranes offers several advantages. Firstly, nanoparticles are typically rigid bodies with well-defined geometry and surface chemistry, allowing them to impose a controlled deformation on the membrane [6]. Additionally, the presence of metals in nanoparticles enhances their visibility in imaging methods. Furthermore, nanoparticles are easier to model compared to complex proteins that can exhibit various conformations [3].

Depending on their properties, nanoparticles can interact with biological membranes in different ways. They may be adsorbed at the membrane surface, translocated through the membrane, or incorporated into the membrane's hydrophobic core [7,8]. In this chapter, our focus will be on the latter interaction, specifically the embedding of hydrophobic nanoinclusions into lipid bilayers. This process shares similarities with the inclusion of integral transmembrane proteins in natural biological membranes. We will begin by describing the properties of nanoparticles that enable their embedding in the membrane. Subsequently, we will discuss the influence of nanoparticles in the membrane's hydrophobic core on its biological functions. Finally, we will explore the mechanisms through which the cell membrane modulates the interactions among spatially distributed nanoparticles.

2. The key role of nanoparticle hydrophobicity

Nanoparticles experience significant van der Waals attractions [9] that lead to irreversible and usually undesirable aggregation [10]. A commonly employed approach to stabilize NPs is through the functionalization of their surface with molecules [11]. These molecules, known as ligands, can be attached to the NP surface either through non-covalent or covalent bonding [12,13]. Non-covalent functionalization relies on a multitude of weak interactions, including electrostatic, ionic, van der Waals, hydrophobic, absorption, and hydrogen bonding, and is particularly suitable for metallic and silica NPs [14,15].

The physicochemical characteristics of the organic ligand shell, which covers the metallic core of NPs, significantly influence their interaction with the biological environment. The use of charged ligands is widespread due to their ability to prevent aggregation in aqueous solutions, which is often a prerequisite for biomedical applications [16]. The behavior of charged NPs depends on the nature and distribution of charges on their surface as well as the charge of the surrounding membrane [17,18]. In a simulation, a NP with a diameter of 3.6 nm was positioned in close proximity to the surface of a bilayer membrane composed of zwitterionic lipids. The neutral NP remained embedded in the pore without significantly affecting the surrounding membrane. However, introducing charges to the NP surface caused local distortion of the bilayer, resulting in a bent configuration (Fig. 1). Furthermore, surface charges facilitated the diffusion of NPs compared to neutral ones [17]. These simulation findings align with experimental observations indicating successful internalization of hydrophobically functionalized gold NPs within the membrane bilayer. Conversely, anionic gold NPs of the same size interacted solely with the outer surface of the membrane [19].

In summary, the stability and behavior of nanoparticles in interaction with biological membranes are intricately intertwined with their surface characteristics and interactions. Specifically, only hydrophobic nanoparticles have the capacity to become embedded within the lipid core of a phospholipid bilayer.

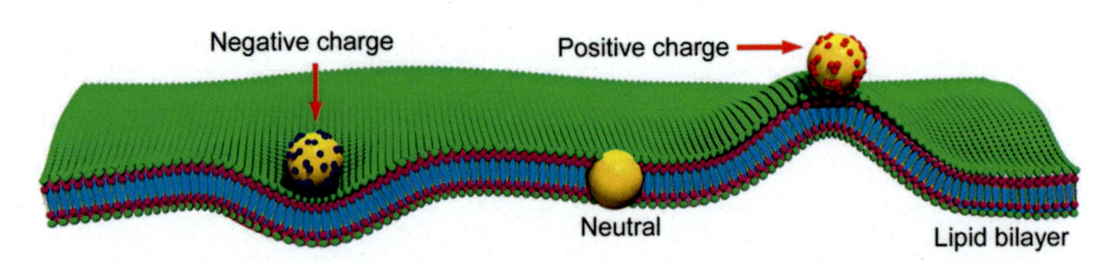

Fig. 1 Interaction of a lipid bilayer membrane with neutral and charged NPs. The phospholipid headgroup contains positively charged choline and negatively charged phosphate. *Adapted from P. Chen, Z. Huang, J. Liang, T. Cui, X. Zhang, B. Miao, et al., Diffusion and directionality of charged nanoparticles on lipid bilayer membrane, ACS Nano 10 (12) (2016) 11541–11547. https://doi.org/10.1021/acsnano.6b07563.*

3. Size matters

Hydrophobic nanoparticle can either be embedded in the nonpolar core of a lipid bilayer or form a micellar structure, as shown in Fig. 2. Sub Wi et al. [20] proposed a simple geometrical model to describe the energy of the biomembrane in both states.

When the NP is embedded within the hydrophobic region of the lipid bilayer, it perturbs the bilayer thickness and induces bending in both bilayer leaflets, as depicted in Fig. 3. The local shape of the lipid bilayer around the NP is influenced by the interplay between the stretching energy of the lipid tails (G_c) and the bending energy of both monolayers (G_b). The bending energy can be described by the deviation of the local membrane curvature (C) from the intrinsic membrane curvature (C_0) [21], while the transversal stretching energy represents the deformation of the hydrophobic core [22]. The total energy is given by:

$$G = \frac{1}{2}\int_A \kappa_b (C - C_0)^2 + \kappa_c \left(\frac{\xi - \xi_0}{\xi_0} \right)^2 \mathrm{d}A \tag{1}$$

In this equation, ξ represents the thickness of the deformed monolayer near the NP, κ_b and κ_c are the bending and compression-expansion moduli of the lipid layers, respectively, and ξ_0 is the length of the lipid tails in an unperturbed monolayer. The integration is performed over the perturbed area.

Sub Wi et al. [20] proposed a model using axisymmetric monolayer deformation profiles to determine the free energy of the lamellar and micellar states of the NP. The shape of both states was parameterized by circular and lamellar contours determined by the radius of the NP (R_{NP}), as shown in Fig. 3.

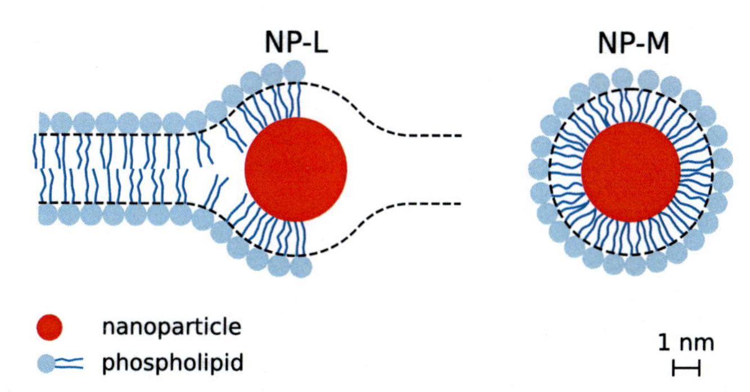

Fig. 2 NP embedded in the lamellar membrane (NP-L) and NP forming a micellar complex (NP-M).

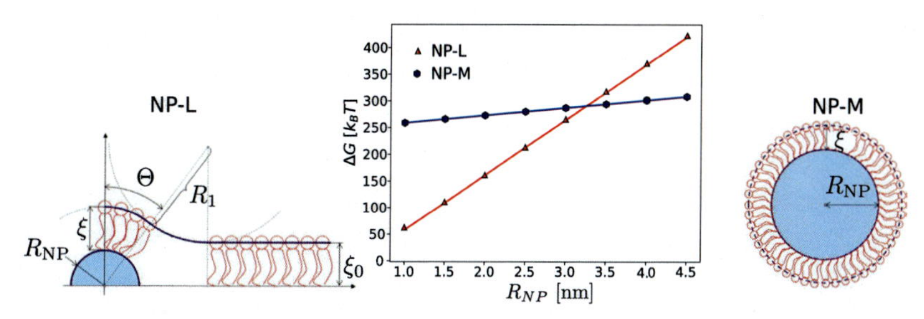

Fig. 3 Schematic geometry of the lamellar state (NP-L) and micellar state (NP-M) of a hydrophobic NP. The elastic deformation energy of NP-L and NP-M with varying NP size is shown. The curves were calculated for DOPC according to the data provided in Sub Wi et al. [20]. *Adapted from H.S. Wi, K. Lee, H.K. Pak, Interfacial energy consideration in the organization of a quantum dot–lipid mixed system, J. Phys. Condens. Matter 20 (49) (2008) 494211. https://doi.org/10.1088/0953-8984/20/49/494211.*

It was discovered that small NPs do not extensively deform the surrounding membrane in the lamellar phase but require significant bending in the micellar phase. Therefore, for small hydrophobic NPs, it is energetically more favorable to be embedded in the hydrophobic core of the phospholipid bilayer. There exists a critical radius, calculated for DOPC as 3.5 nm, beyond which the NP is not stable in the lamellar phase but prefers the energetically favorable micellar phase [20]. Similar results were obtained assuming a lipid packing model, as shown in Sung et al. [23].

The existence of a critical radius has been confirmed in experiments with NPs of various sizes. In a study involving hydrophobic quantum dots, it was found that blue, green, and yellow-emitting quantum dots (with core radii of 1.05 nm, 1.25 nm, and 1.65 nm, respectively) were incorporated into an egg phospholipid membrane, while red-emitting quantum dots (with a core radius of 2.5 nm) were not [23].

The effect of NP size on the interaction with biomembranes was further investigated using coarse-grained simulations [24]. NPs with hydrophobic functionalization and diameters of 2 nm, 4 nm, and 8 nm were modeled. It was observed that NP intake and release were associated with energy barriers. Intake of an 8 nm nanoparticle required substantial deformation of the lipid bilayer, resulting in a repulsion force of 0.2 nN and an energy barrier of 40 k_BT. The energy barrier was considerably higher than those for 2 nm and 4 nm NPs (3.4 k_BT and 4.2 k_BT, respectively). The release energy barriers were significantly higher than the intake barriers (75 k_BT, 37 k_BT, and 110 k_BT for NPs with radii of 2 nm, 4 nm, and 8 nm, respectively). The release energy barrier for the 4 nm NP was smaller than those for the 2 nm and 8 nm NPs, implying the existence of an optimal NP size for unforced transmembrane transport [24].

The results of the simple model are also in agreement with predictions based on the single-chain mean field (SCMF) theory. Strong hydrophobic interactions between the membrane core and the hydrophobic particle lead to a structural rearrangement of lipids near the NP. Small NPs with diameters of 5 nm or less can be fully inserted into the bilayer without significant disruption, while larger NPs with diameters greater than 5 nm alter the bilayer structure and are likely to exist in the lamellar form [25].

Although hydrophobic NPs larger than 6 nm tend to induce surface absorption rather than embedding into the lipid bilayer [26], this does not hold true for clusters of small NPs. Experimental studies have shown that superparamagnetic iron oxide NPs with a diameter of 6 nm, stabilized by β-octyl glucoside, form flexible structures between two monolayers up to 60 nm in size [27]. Dendrimers also exhibit similar flexible aggregates in the lipid bilayer [28].

In summary, when a hydrophobic NP becomes incorporated into the hydrophobic region of the lipid bilayer, it triggers perturbations in the bilayer's thickness and induces bending in both leaflets. Smaller hydrophobic NPs find it energetically advantageous to embed themselves within the core of the phospholipid bilayer. Conversely, larger NPs encounter significant energy barriers during the processes of intake and release.

4. Nanoparticles-mediated effect on the membrane

4.1 Phase behavior and fluidity

Lipids exhibit two primary phases: the gel phase and the liquid phase. The temperature, impacting the free energy of membrane constituents, governs the shift from the gel phase to the liquid phase when temperatures surpass a lipid's T_m (melting temperature) [29]. However, various factors that influence lipid organization can also sway T_m. Upon the introduction of hydrophobic NPs into the membrane, these nanoparticles engage with phospholipid tails, subsequently affecting their arrangement. One might anticipate that the perturbation of lipid packing, induced by the curvature resulting from nanoparticles, coupled with the interaction between hydrophobic chains and the nanoparticles, will produce measurable consequences on T_m. Despite considerable research into the impact of nanoparticles on membrane thermodynamics, the findings have yielded mixed interpretations.

It has been found that Ag-decanethiol NPs with a diameter of 6 nm incorporated into DPPC and DPPS membranes have little effect on the transition temperature at low concentrations [30]. However, there is a decrease in the melting temperature of 5 °C at DPPC/AgNP ratios less than 15:1. The AgNPs also increase the fluidity of the bilayer, which means they reduce lipid ordering in the gel phase.

On the contrary, stabilized 5 nm CdSe quantum dots [31] and decanthiol-stabilized 5.7 nm Ag NPs [30] did not influence T_m when embedded in zwitterionic membranes. In contrast, oleic acid-coated iron oxide NPs embedded in zwitterionic liposome membranes have been reported to increase T_m, whereas stabilized CdSe/ZnS quantum dots decreased and broadened T_m of cationic liposomes [32].

Bilayer-embedded 5 nm oleic acid-capped superparamagnetic iron oxide NPs led to clear changes in biomembrane structure, lipid phase behavior, and transbilayer leakage [33]. With embedded superparamagnetic iron oxide NPs, DPPC exhibited suppressed pretransitions and merged with lipid melting at higher temperatures [33]. Additionally, the incorporation of NPs decreases the inherent leak of phospholipid vesicles.

Fulleren NPs (C60) trapped within the vesicle bilayer led to a reduction in the lipid melting temperature [34]. Fluorescence anisotropy results indicate that bilayer-embedded C60 increases the degree of lipid ordering by inhibiting lipid motion and increasing ordering, which is in agreement with theoretical predictions [35]. X-ray reflectivity profiles for oriented

DPPC/C60 bilayers indicate a larger head-to-head distance compared to pure DPPC [36]. Small-angle neutron scattering observed in-plane ripple-like ordering together with deteriorated normal-to-plane ordering [36].

In addition to NPs, the hydrophobic core of the biomembrane may serve as a solvent for other hydrophobic polymers [37]. In a computational study, the interaction of polymer NPs made of polyethylene (PE), polypropylene (PP), and polystyrene (PS) with sizes up to 7 nm with the membrane was studied. While PE shows a strong tendency to self-aggregate within lipid bilayers, resembling solid NPs, PP and PS dissolve completely. The inclusion of PE NPs into the cell membrane affects its order locally. For instance, the membrane thickness is strongly increased in the regions where the PE aggregates, and the order parameter of lipid chains is decreased due to membrane bending and possibly higher disorder (Fig. 4). It has even been shown that the inclusion of PE NPs results in lipid separation for ternary lipid mixtures [37].

It was found that spherical gold NPs with a diameter of 3–4 nm entrapped inside the DPPC bilayer increase the fluidity of the bilayer [38]. Below the transition temperature, gold NPs have little effect on membrane fluidity. However, above the transition temperature, as the concentration of gold NPs in the membrane increases, the membrane fluidities also increase [38]. However, this finding has not been confirmed in the study by Santhosh et al., where the incorporation of 2 nm gold NPs at low concentrations of up to 2% wt did not show any notable changes in membrane fluidity [39].

The inclusion of NPs also has an effect on the electrical properties of the membrane. Basham et al. showed that vesicles of DOPC loaded with hydrophobic NPs with a diameter of 3 nm exhibit higher ion conductance [40].

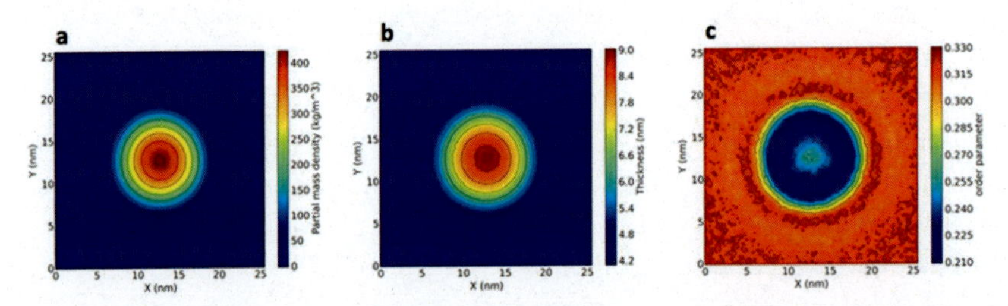

Fig. 4 Molecular simulation of polyethylene NP (PE80) in the POPC lipid bilayer: (A) PE density, (B) membrane thickness, and (C) lipid chains' order parameter. *Adapted from D. Bochicchio, E. Panizon, L. Monticelli, G. Rossi, Interaction of hydrophobic polymers with model lipid bilayers, Sci. Rep. 7 (1) (2017) 6357. https://doi.org/10.1038/s41598-017-06668-0.*

In addition, entrapped NPs increased membrane tension and reduced the energy of membrane adhesion. Cation transport across phospholipid membranes can also be enhanced by ligand-capped gold NPs [41].

4.2 NPs induce membrane softening

The change in the membrane phase state is linked with the membrane mechanics as predicted by Heimburg [29]. Neutron spin spectroscopy demonstrated approximately 15% decrease in the bending modulus for liposomes loaded with small hydrophobic gold nanoparticles (with diameters of 3 nm and 3.8 nm) relative to pure liposomes [42]. The bending modulus κ_b is related to the area compressibility modulus κ_a as follows [43]:

$$\kappa_a = \beta \frac{\kappa}{(2h_c)^2} \tag{2}$$

where β is a coupling constant that depends on the degree of coupling between the bilayers, ranging from 12 (strong coupling) to 48 (no coupling), and h_c is the monolayer thickness. There is not enough information about the effect of nanoparticle inclusion on the coupling constant β. However, the inclusion of nanoparticles in the hydrophobic membrane core does not change the bilayer thickness [42]. A decrease in the bending modulus without a change in membrane thickness indicates a decrease in the area compressibility modulus. A similar decrease in the membrane bending and area compressibility modulus can be achieved by the inclusion of molecules into the biomembrane that decrease the energy of interaction between phospholipids and increase the mean surface occupied by each phospholipid molecule [44]. An increase in the bilayer area without changing the number of lipids could be related to the conformation change of lipid tails, i.e., the transition from the gel to fluid phase [29]. Liposomes formed with gold nanoparticles of diameter 4 nm in a microfluidic device also show a considerable decrease in the area compressibility modulus for POPC, from a maximum value of 130 mN/m for the pure bilayer to 30 mN/m for POPC with nanoparticles [45].

Interestingly, liposomes containing small iron oxide nanoparticles in their membranes are significantly smaller than the respective reference liposomes [46]. A similar observation concerning the decreased size of extruded polymersomes after the incorporation of hydrophobic quantum dots into the membrane was made by Binder et al. in 2007 [47]. The reason for the reduced size could be membrane wrapping induced by the incorporated nanoparticles [48].

In summary, when NPs are embedded within the lipid bilayer, they influence membrane fluidity and consequently impact the transition temperature. However, there isn't a singular, consistent effect of nanoparticles on the transition temperature; instead, it's probable that NPs exert their influence by counteracting various effects. For instance, increased curvature might disrupt lipid packing, yet strong hydrophobic interactions could lead to enhanced lipid packing. The final outcome will depend on the dominant mechanism at play.

5. Membrane-mediated effects on NP aggregation

As shown in the previous section, there is an inherent link between the properties of nanoparticles and the behavior of membranes. Interactions between inclusions at membranes can classified into two main categories: direct and indirect forces [49]. Direct interactions represent chemical and physical forces between particles and biomolecules, such as van der Waals or electrostatic forces. On the other hand, indirect or membrane-mediated interactions are generated by a physical perturbation applied to the membrane bilayer structure or its shape by the attached/embedded inclusions [50].

Indirect membrane-mediated interactions include those induced by membrane tension, membrane thickness perturbation, or perturbing membrane curvature [51–53]. When nanoparticles interact with the membrane, they perturb the membrane, causing short-range and long-range forces that result in either their clustering (attractive forces) or dissolving in the membrane (repulsive forces) [52,54,55]. Within this view, the membrane allows the interaction between spatially separated nanoparticles, meaning the membrane can be perceived as a field transferring forces between distant nanoparticles [56].

Short-range membrane-mediated interactions between nanoparticles are usually related to the perturbation of the packing of nearby lipid chains, similar to proteins [57]. Coarse-grained molecular simulations showed that the affected region vanishes at a distance of approximately 2.5 nm from the center of a nanoparticle with a 10 nm diameter [58]. Molecular dynamics simulations showed that embedded nanoparticles locally thin the bilayer and induce negative hydrophobic mismatch [57]. The negative hydrophobic mismatch is known to be related to protein clustering [59].

There are two types of long-range interactions between protein inclusions mediated by the embedding fluid membranes: fluctuation-induced interactions and curvature-induced elastic interactions. Fluctuation-induced interactions originate from the modification of membrane fluctuations, while curvature-induced elastic interactions result from the perturbation of the equilibrium membrane shape by the presence of inclusions [60].

Asymmetrically inserted hydrophobic particles [35] might form an intrinsic curvature of the bilayer [61]. The local influence of membrane curvature has been extensively studied in membranes, showing that such inclusions could generate attraction or repulsion based on inclusion intrinsic curvature and membrane properties [62]. It has been further found that amphiphilic Au nanoparticles with a diameter of 4 nm have the potential to generate curvature in phosphatidylcholine lipid membranes [63].

Molecular simulations of POPC lipid membranes loaded with spherical hydrophobic nanoparticles show that the aggregation of particles depends on their size [64]. The smallest nanoparticles (1.13 nm diameter) show little tendency to aggregate, while nanoparticles larger than the membrane thickness (4.63 nm diameter) form large aggregates. Middle-sized nanoparticles tend to form linear aggregates. Linear aggregation occurs preferentially at regions with higher curvature, and the mechanisms of linear aggregate formation are driven by membrane undulation. Suppressing membrane undulations significantly reduces linear aggregation. It has been proposed that even spherically symmetric nanoparticles can both sense membrane curvature (at low nanoparticle concentration) and induce it (at high nanoparticle concentration), similar to proteins [64].

The unique properties of gold quantum dots allow for the direct monitoring of nanoparticle aggregation [65]. A decrease in the interparticle distance after aggregation in the biological membrane leads to the appearance of a secondary peak centered at about 610 nm, which is associated with a color change of the dispersion from red to purple or dark blue [66]. This red-shift is caused by plasmon-plasmon coupling and indicates the membrane-templated aggregation of nanoparticles [67,68]. Caselli et al. (2021) showed using charged citrated gold nanoparticles with a diameter of 13 nm interacting with various biomembranes (DSPC, DPPC/DSPC, DPPC, POPC/DPPC, POPC, DOPC) that the stiffness of the target vesicle finely modulates the extent of AuNP aggregation, which can be monitored by UV-Vis spectrophotometry [68]. In particular, the intensity of the red-shifted peak, i.e., the hallmark of AuNP aggregation, is minimized for rigid liposomes enveloped by a gel-phase bilayer. On the contrary, the lipid membrane-nanoparticle interaction and, in turn, the red-shifted peak are maximized for soft liposomes with liquid crystalline membranes, which are able to efficiently bend and wrap the nanoparticles [69]. Although larger nanoparticles might be attached to the surface of lipid vesicles, the phenomenon was further studied with smaller nanoparticles [66]. Molecular dynamics simulation indicates that fluid-phase DOPC bilayers completely wrap the nanoparticles, while nanoparticles with gel biomembranes are only partially integrated and form a bridge between adjacent bilayers [66]. However, in a similar study of Ag nanoparticles, it was shown that the characteristic surface plasmon resonance wavelength of the embedded nanoparticles is independent of the bilayer phase [30] (Fig. 5).

Tian et al. studied the aggregation mechanism of hydrophobic nanoparticles (NPs) inside a lipid membrane using a coarse-grained model [70]. They found that hydrophobic NPs can self-assemble into clusters of distinct morphologies or remain in an unclustered state, depending mainly on the NP size and membrane tension. An increase in lateral isotropic tension induces the NP to transition from an unclustered configuration to aggregation into clusters or forming linear segments. The existence of NP clusters also depends on the size of the NP, with larger NPs exhibiting clusters at smaller lateral tension. Aggregation of NPs was observed when the NP size was comparable to the size of the lipid bilayer, while no cluster formation was observed for NP diameters less than 3 nm or larger than 12.0 nm, with NPs assuming a gas-like state in the hydrophobic core of the lipid bilayer.

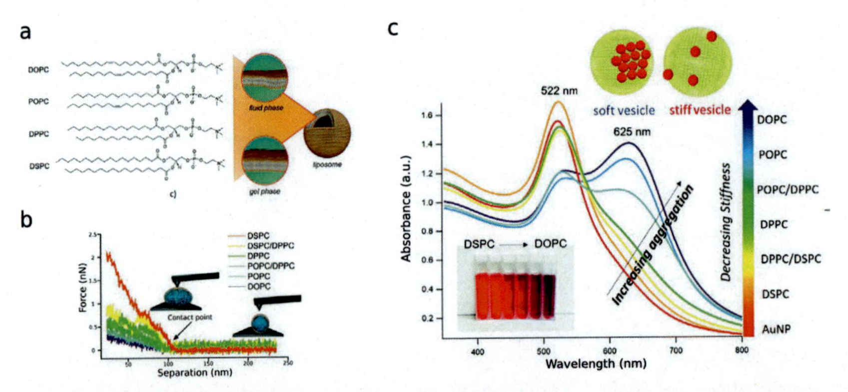

Fig. 5 (A) Phase and chemical formulas of the four lipids used for the preparation of liposomes, (B) AFM force-distance curves, and (C) UV-Vis spectra and visual appearance of the samples of gold NPs incubated with synthetic vesicles. *Adopted from L. Caselli, A. Ridolfi, J. Cardellini, L. Sharpnack, L. Paolini, M. Brucale, et al., A plasmon-based nanoruler to probe the mechanical properties of synthetic and biogenic nanosized lipid vesicles, Nanoscale Horiz. 6 (7) (2021) 543–550. https://doi.org/10.1039/D1NH00012H.*

Fig. 6 illustrates clusters of NPs in the hydrophobic core of the lipid bilayer under varying membrane tension. The membrane tension increases from left to right, with the NP diameter set to 5.0 nm. At intermediate values of membrane tension, the formation of NP clusters is observed.

The physical fields affecting the biological membrane also have an impact on embedded NPs. It was shown that by increasing the transmembrane voltage on a DOPC membrane loaded with gold NPs functionalized with hydrophobic octanethiol, the clustered NPs become separated [40]. The applied voltage exerts an electrocompressive force on the membrane that acts to disperse NP clusters. A critical voltage of 150 mV was observed, at which the bilayer thickness suddenly decreased, indicating a change in the state of the NPs. This effect was accompanied by a drop in capacitance at a constant area, implying a thickening of the membrane's hydrophobic core.

Membrane-mediated interactions can also be studied in artificial systems [6]. Lamellar phases of a zwitterionic surfactant doped with monodisperse and spherical hydrophobic inorganic particles showed the formation of a dense layer of NPs in the hydrophobic membrane core. Analysis of synchrotron small-angle x-ray scattering indicated repulsive forces between NPs induced by the membrane, with a contact value of about 4 k_BT and a range of 14 Å.

The method of NP loading into lipid vesicles can also affect NP aggregation [71]. When phosphatidylcholine and NP lipid vesicles are prepared by extrusion, the NPs form a dense monolayer in the

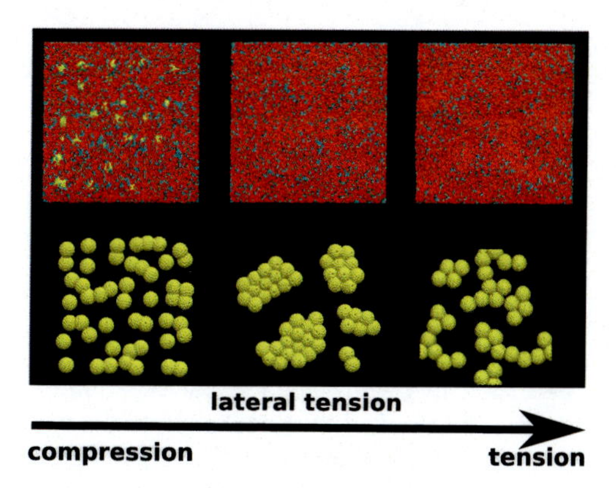

Fig. 6 Clusters of NPs in the hydrophobic core of the lipid bilayer. The membrane tension increases from left (compression) to right (tension). The membrane tension is close to zero in the middle. *Adapted from F. Tian, T. Yue, Y. Li, X. Zhang, Computer simulation studies on the interactions between nanoparticles and cell membrane, Sci. China Chem. 57 (12) (2014) 1662–1671. https://doi.org/10.1007/s11426-014-5231-7.*

hydrophobic core of the lipid bilayer. However, if pure phospholipid vesicles are prepared first and then loaded with NPs using the dialysis method, the membrane contains regions with and without NPs (Fig. 7). In these vesicles, called Janus-like vesicles, the NPs aggregate and form clusters in the NP-rich membrane regions.

Daniel et al. studied the energy of NP clustering by considering bending and contact energy [72]. They found that the two stable states of NPs in the vesicle membrane can be explained by the existence of an energy barrier between aggregated and separated NPs mediated by the biomembrane [71]. Analysis of the membrane's mechanical parameters revealed that the energy barrier between two membrane-embedded NPs is mainly due to bending deformation rather than a change in the thickness of the bilayer in the vicinity of the nanoparticle [72].

In summary, NPs embedded within lipid bilayers interact with each other through the lipid membrane, which serves as a medium for force transmission. These interactions result in the clustering (attractive forces) or dispersion (repulsive forces) of NPs within the membrane, contingent upon the characteristics of both the membrane and the nanoparticles. The nature of these membrane-mediated interactions is in addition to NP characteristics influenced by factors such as membrane composition and environmental conditions like temperature, lateral tension, and transmembrane voltage.

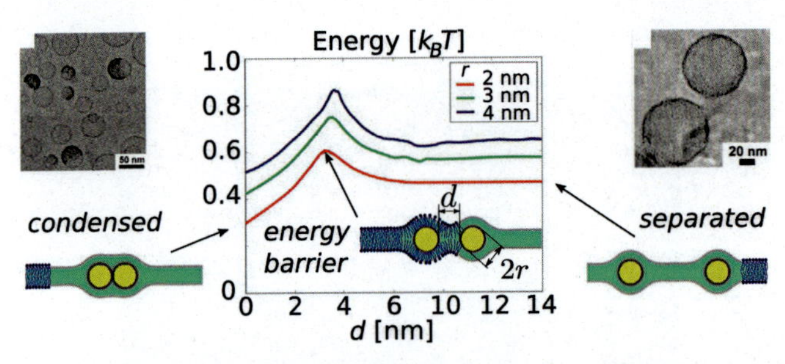

Fig. 7 (Left) Janus-type NP-vesicle hybrids prepared by dialysis of a suspension of vesicles and detergent-dispersed NPs, (Right) vesicles prepared by extrusion in the presence of NPs, where NPs in the membrane are separated. (Middle) Predicted energy barrier between the clustered and separated states of nanoparticles. *From M.R. Rasch, E. Rossinyol, J.L. Hueso, B.W. Goodfellow, J. Arbiol, B.A. Korgel, Hydrophobic gold nanoparticle self-assembly with phosphatidylcholine lipid: membrane-loaded and janus vesicles, Nano Lett. 10 (9) (2010) 3733–3739. https://doi.org/10.1021/nl102387n and M. Daniel, J. Řezníčková, M. Handl, A. Iglič, V. Kralj-Iglič, Clustering and separation of hydrophobic nanoparticles in lipid bilayer explained by membrane mechanics, Sci. Rep. 8 (1) (2018) 10810. https://doi.org/10.1038/s41598-018-28965-y, with permission.*

6. Conclusions

The ability of nanoparticles (NPs) to fully embed in the lipid bilayer provides valuable insights into the effect of bilayer inclusions on membrane function. Understanding the mutual influence between NPs and the membrane is essential for describing such interactions. While the thickness of the hydrophobic moiety primarily determines the critical size of lipid NPs, the interaction between NPs embedded in the lipid bilayer depends on how the NPs affect the membrane. The study of bilayer-NP interactions has been approached through various theoretical methods [16] and biomimetic membrane models [73]. It was shown that the interaction between the membrane and nanoparticles can lead to changes in the membrane state, such as alterations in fluidity, stiffness, or curvature, as well as the aggregation of embedded nanoparticles. NP aggregation within lipid membranes is a complex process influenced by factors such as NP size, membrane tension, external physical fields, and even the methods used for NP/biomembrane preparation.

The mutual interaction of nanoparticle-membrane composites holds promise in the development of aqueous-based sensors and provides a model system for studying NP/cell interactions. Nanoparticles can directly influence protein function by affecting the membrane state [74]. They can also be utilized to initiate and control drug release in response to external stimuli [26,75–77]. Magnetic NPs offer the advantage of selective release of encapsulated molecules through NP heating using alternating current electromagnetic fields [46]. Laser activation can also be employed, as metal NPs can absorb light and transiently change the membrane permeability in their immediate vicinity [78]. Changes in the external environment, such as pH, can also affect NP/membrane complexes and lead to the development of novel pH-sensitive drug delivery systems [79,80]. Another application of membrane-NP integration lies in diagnostic cellular imaging. NPs can enhance fluorescence and act as contrast agents [81] and even detect transmembrane potentials [40].

References

[1] G.L. Nicolson, The fluid-mosaic model of membrane structure: still relevant to understanding the structure, function and dynamics of biological membranes after more than 40 years, Biochim. Biophys. Acta 1838 (6) (2014) 1451–1466, https://doi.org/10.1016/j.bbamem.2013.10.019

[2] M.F. Brown, Curvature forces in membrane lipid-protein interactions, Biochemistry 51 (49) (2012) 9782–9795, https://doi.org/10.1021/bi301332v

[3] V. Corradi, B.I. Sejdiu, H. Mesa-Galloso, H. Abdizadeh, S.Y. Noskov, S.J. Marrink, et al., Emerging diversity in lipid–protein interactions, Chem. Rev. 119 (9) (2019) 5775–5848, https://doi.org/10.1021/acs.chemrev.8b00451

[4] E.P. Carpenter, K. Beis, A.D. Cameron, S. Iwata, Overcoming the challenges of membrane protein crystallography, Curr. Opin. Struct. Biol. 18 (5) (2008) 581–586, https://doi.org/10.1016/j.sbi.2008.07.001
[5] M.P. Muller, T. Jiang, C. Sun, M. Lihan, S. Pant, P. Mahinthichaichan, et al., Characterization of lipid-protein interactions and lipid-mediated modulation of membrane protein function through molecular simulation, Chem. Rev. 119 (9) (2019) 6086–6161, https://doi.org/10.1021/acs.chemrev.8b00608
[6] D. Constantin, B. Pansu, M. Impéror, P. Davidson, F. Ribot, Repulsion between inorganic particles inserted within surfactant bilayers, Phys. Rev. Lett. 101 (9) (2008) 098101, https://doi.org/10.1103/PhysRevLett.101.098101
[7] J. Agudo-Canalejo, R. Lipowsky, Particle–membrane interactions, [2020] ISBN 978-1-315-15251-6, in: R. Dimova, C.M. Marques (Eds.), The Giant Vesicle Book, first ed., CRC Press, Taylor & Francis Group, Boca Raton, FL, 2019, pp. 211–227, https://doi.org/10.1201/9781315152516-8
[8] E. Okoampah, Y. Mao, S. Yang, S. Sun, C. Zhou, Gold nanoparticles–biomembrane interactions: from fundamental to simulation, Colloids Surfaces B: Biointerfaces 196 (2020) 111312, https://doi.org/10.1016/j.colsurfb.2020.111312
[9] E. Alipour, D. Halverson, S. McWhirter, G.C. Walker, Phospholipid bilayers: stability and encapsulation of nanoparticles, Annu. Rev. Phys. Chem. 68 (1) (2017) 261–283, https://doi.org/10.1146/annurev-physchem-040215-112634
[10] R. Zangi, B.J. Berne, Aggregation and dispersion of small hydrophobic particles in aqueous electrolyte solutions, J. Phys. Chem. B 110 (45) (2006) 22736–22741, https://doi.org/10.1021/jp064475+
[11] G. Sanità, B. Carrese, A. Lamberti, Nanoparticle surface functionalization: how to improve biocompatibility and cellular internalization, Front. Mol. Biosci. 7 (2020).
[12] M.J. Rybak-Smith, Effect of surface modification on toxicity of nanoparticles, in: B. Bhushan (Ed.), Encyclopedia of Nanotechnology, Springer, Netherlands, Dordrecht, 2012, pp. 645–652, https://doi.org/10.1007/978-90-481-9751-4_174
[13] K. Skłodowski, S.J. Chmielewska-Deptuła, E. Piktel, P. Wolak, T. Wollny, R. Bucki, Metallic nanosystems in the development of antimicrobial strategies with high antimicrobial activity and high biocompatibility, Int. J. Mol. Sci. 24 (3) (2023) 2104, https://doi.org/10.3390/ijms24032104
[14] C. Contini, M. Schneemilch, S. Gaisford, N. Quirke, Nanoparticle–membrane interactions, J. Exp. Nanosci. 13 (2017) 1–20, https://doi.org/10.1080/17458080.2017.1413253
[15] P. Singh, S. Pandit, V.R.S.S. Mokkapati, A. Garg, V. Ravikumar, I. Mijakovic, Gold nanoparticles in diagnostics and therapeutics for human cancer, Int. J. Mol. Sci. 19 (7) (2018) 1979, https://doi.org/10.3390/ijms19071979
[16] G. Rossi, L. Monticelli, Gold nanoparticles in model biological membranes: a computational perspective, Biochim. Biophys. Acta – Biomembr. 1858 (10) (2016) 2380–2389, https://doi.org/10.1016/j.bbamem.2016.04.001
[17] P. Chen, Z. Huang, J. Liang, T. Cui, X. Zhang, B. Miao, et al., Diffusion and directionality of charged nanoparticles on lipid bilayer membrane, ACS Nano 10 (12) (2016) 11541–11547, https://doi.org/10.1021/acsnano.6b07563
[18] C.-F. Su, H. Merlitz, H. Rabbel, J.-U. Sommer, Nanoparticles of various degrees of hydrophobicity interacting with lipid membranes, J. Phys. Chem. Lett. 8 (17) (2017) 4069–4076, https://doi.org/10.1021/acs.jpclett.7b01888
[19] S. Tatur, M. Maccarini, R. Barker, A. Nelson, G. Fragneto, Effect of functionalized gold nanoparticles on floating lipid bilayers, Langmuir 29 (22) (2013) 6606–6614, https://doi.org/10.1021/la401074y
[20] H.S. Wi, K. Lee, H.K. Pak, Interfacial energy consideration in the organization of a quantum dot–lipid mixed system, J. Phys. Condens. Matter 20 (49) (2008) 494211, https://doi.org/10.1088/0953-8984/20/49/494211

[21] W. Helfrich, Elastic properties of lipid bilayers: theory and possible experiments, Z. Für Naturforschung Tl C Biochem. Biophys. Biol. Virol 11 (28) (1973) 693–703.
[22] Š. Perutková, M. Daniel, M. Rappolt, G. Pabst, G. Dolinar, V. Kralj-Iglič, et al., Elastic deformations in hexagonal phases studied by small-angle X-ray diffraction and simulations, Phys. Chem. Chem. Phys. 13 (8) (2011) 3100–3107, https://doi.org/10.1039/c0cp01187h
[23] S.K. Sung, H.K. Pak, J.H. Kwak, S.W. Lee, Y.H. Kim, B.I. Hur, et al., Specific radius change of quantum dot inside the lipid bilayer by charge effect of lipid head-group, Open J. Biophys. 8 (3) (2018) 163–175, https://doi.org/10.4236/ojbiphy.2018.83012
[24] S. Burgess, Z. Wang, A. Vishnyakov, A.V. Neimark, Adhesion, intake, and release of nanoparticles by lipid bilayers, J. Colloid Interface Sci. 561 (2020) 58–70, https://doi.org/10.1016/j.jcis.2019.11.106
[25] Y. Guo, E. Terazzi, R. Seemann, J.B. Fleury, V.A. Baulin, Direct proof of spontaneous translocation of lipid-covered hydrophobic nanoparticles through a phospholipid bilayer, Sci. Adv. 2 (11) (2016) 38–40, https://doi.org/10.1126/sciadv.1600261
[26] M.R. Preiss, G.D. Bothun, Stimuli-responsive liposome-nanoparticle assemblies, Expert Opin Drug Deliv. 8 (8) (2011) 1025–1040, https://doi.org/10.1517/17425247.2011.584868
[27] C. Bonnaud, C.A. Monnier, D. Demurtas, C. Jud, D. Vanhecke, X. Montet, et al., Insertion of nanoparticle clusters into vesicle bilayers, ACS Nano 8 (4) (2014) 3451–3460, https://doi.org/10.1021/nn406349z
[28] R. Guo, J. Mao, L.-T. Yan, Unique dynamical approach of fully wrapping dendrimer-like soft nanoparticles by lipid bilayer membrane, ACS Nano 7 (12) (2013) 10646–10653, https://doi.org/10.1021/nn4033344
[29] T. Heimburg, Mechanical aspects of membrane thermodynamics. Estimation of the mechanical properties of lipid membranes close to the chain melting transition from calorimetry, Biochim. Biophys. Acta (BBA) – Biomembranes 1415 (1) (1998) 147–162, https://doi.org/10.1016/S0005-2736(98)00189-8
[30] G.D. Bothun, Hydrophobic silver nanoparticles trapped in lipid bilayers: size distribution, bilayer phase behavior, and optical properties, J. Nanobiotechnol. 6 (2008) 13, https://doi.org/10.1186/1477-3155-6-13
[31] G. Gopalakrishnan, C. Danelon, P. Izewska, M. Prummer, P.-Y. Bolinger, I. Geissbühler, et al., Multifunctional lipid/quantum dot hybrid nanocontainers for controlled targeting of live cells, Angew. Chem. Int. Ed. Engl. 45 (33) (2006) 5478–5483, https://doi.org/10.1002/anie.200600545
[32] G.D. Bothun, A.E. Rabideau, M.A. Stoner, Hepatoma cell uptake of cationic multifluorescent quantum dot liposomes, J. Phys. Chem. B 113 (22) (2009) 7725–7728, https://doi.org/10.1021/jp9017458
[33] Y. Chen, A. Bose, G.D. Bothun, Controlled release from bilayer-decorated magnetoliposomes via electromagnetic heating, ACS Nano 4 (6) (2010) 3215–3221, https://doi.org/10.1021/nn100274v
[34] Y. Chen, G.D. Bothun, Lipid-assisted formation and dispersion of aqueous and bilayer-embedded nano-C60, Langmuir 25 (9) (2009) 4875–4879, https://doi.org/10.1021/la804124q
[35] R. Qiao, A.P. Roberts, A.S. Mount, S.J. Klaine, P.C. Ke, Translocation of C60 and its derivatives across a lipid bilayer, Nano Lett. 7 (3) (2007) 614–619, https://doi.org/10.1021/nl062515f
[36] U.-S. Jeng, C.-H. Hsu, T.-L. Lin, C.-M. Wu, H.-L. Chen, L.-A. Tai, et al., Dispersion of fullerenes in phospholipid bilayers and the subsequent phase changes in the host bilayers, Phys. B: Condens. Matter 357 (1) (2005) 193–198, https://doi.org/10.1016/j.physb.2004.11.055

[37] D. Bochicchio, E. Panizon, L. Monticelli, G. Rossi, Interaction of hydrophobic polymers with model lipid bilayers, Sci. Rep. 7 (1) (2017) 6357, https://doi.org/10.1038/s41598-017-06668-0
[38] S.-H. Park, S.-G. Oh, J.-Y. Mun, S.-S. Han, Loading of gold nanoparticles inside the DPPC bilayers of liposome and their effects on membrane fluidities, Colloids and Surfaces B: Biointerfaces 48 (2) (2006) 112–118, https://doi.org/10.1016/j.colsurfb.2006.01.006
[39] P.B. Santhosh, T. Tenev, L. Šturm, N.P. Ulrih, J. Genova, Effects of hydrophobic gold nanoparticles on structure and fluidity of SOPC lipid membranes, Int. J. Mol. Sci. 24 (12) (2023) 10226, https://doi.org/10.3390/ijms241210226
[40] C.M. Basham, S. Spittle, J. Sangoro, J. El-Beyrouthy, E. Freeman, S.A. Sarles, Entrapment and voltage-driven reorganization of hydrophobic nanoparticles in planar phospholipid bilayers, ACS Appl. Mater. Interfaces 14 (49) (2022) 54558–54571, https://doi.org/10.1021/acsami.2c16677
[41] M.P. Grzelczak, S.P. Danks, R.C. Klipp, D. Belic, A. Zaulet, C. Kunstmann-Olsen, et al., Ion transport across biological membranes by carborane-capped gold nanoparticles, ACS Nano 11 (12) (2017) 12492–12499, https://doi.org/10.1021/acsnano.7b06600
[42] S. Chakraborty, A. Abbasi, G.D. Bothun, M. Nagao, C.L. Kitchens, Phospholipid bilayer softening due to hydrophobic gold nanoparticle inclusions, Langmuir 34 (44) (2018) 13416–13425, https://doi.org/10.1021/acs.langmuir.8b02553
[43] D. Boal, Mechanics of the cell, Africa (Lond), Cambridge University Press, Cambridge, 2002, https://doi.org/10.1016/S0092-8674(02)00789-4
[44] N. Fa, L. Lins, P.J. Courtoy, Y. Dufrêne, P. Van Der Smissen, R. Brasseur, et al., Decrease of elastic moduli of DOPC bilayers induced by a macrolide antibiotic, azithromycin, Biochim. Biophys. Acta (BBA) – Biomembranes 1768 (7) (2007) 1830–1838, https://doi.org/10.1016/j.bbamem.2007.04.013
[45] J. Perrotton, R. Ahijado-Guzmán, L.H. Moleiro, B. Tinao, A. Guerrero-Martinez, E. Amstad, et al., Microfluidic fabrication of vesicles with hybrid lipid/nanoparticle bilayer membranes, Soft Matter 15 (6) (2019) 1388–1395, https://doi.org/10.1039/C8SM02050G
[46] E. Amstad, J. Kohlbrecher, E. Müller, T. Schweizer, M. Textor, E. Reimhult, Triggered release from liposomes through magnetic actuation of iron oxide nanoparticle containing membranes, Nano Lett. 11 (4) (2011) 1664–1670, https://doi.org/10.1021/nl2001499
[47] W.H. Binder, R. Sachsenhofer, D. Farnik, D. Blaas, Guiding the location of nanoparticles into vesicular structures: a morphological study, Phys. Chem. Chem. Phys. 9 (48) (2007) 6435–6441, https://doi.org/10.1039/B711470M
[48] C. Contini, J.W. Hindley, T.J. Macdonald, J.D. Barritt, O. Ces, N. Quirke, Size dependency of gold nanoparticles interacting with model membranes, Commun. Chem. 3 (1) (2020) 1–12, https://doi.org/10.1038/s42004-020-00377-y
[49] E. Alizadeh-Haghighi, A. Karaei Shiraz, A.H. Bahrami, Membrane-mediated interactions between disk-like inclusions adsorbed on vesicles, Front. Phys. 10 (2022).
[50] A.-F. Bitbol, D. Constantin, J.-B. Fournier, Membrane-mediated interactions, Physics of Biological Membranes, Springer, 2018, pp. 311–350, https://doi.org/10.1007/978-3-030-00630-3_13
[51] M. Daniel, K. EleršičFilipič, E. Filová, T. Judl, J. Fojt, Modelling the role of membrane mechanics in cell adhesion on titanium oxide nanotubes, Comput. Methods Biomech. Biomed. Eng. (2022) 1–10, https://doi.org/10.1080/10255842.2022.2058875
[52] Y. Jiang, B. Thienpont, V. Sapuru, R.K. Hite, J.S. Dittman, J.N. Sturgis, et al., Membrane-mediated protein interactions drive membrane protein organization, Nat. Commun. 13 (1) (2022) 7373, https://doi.org/10.1038/s41467-022-35202-8

[53] J. Midya, T. Auth, G. Gompper, Membrane-mediated interactions between non-spherical elastic particles, ACS Nano 17 (3) (2023) 1935–1945, https://doi.org/10.1021/acsnano.2c05801

[54] H. Alimohamadi, P. Rangamani, Modeling membrane curvature generation due to Membrane‾ Protein interactions, Biomolecules 8 (4) (2018) 35–38, https://doi.org/10.3390/biom8040120

[55] A.F. Bitbol, P.G. Dommersnes, J.B. Fournier, Fluctuations of the Casimir-like force between two membrane inclusions, Phys. Rev. E – Stat. Nonlinear Soft Matter Phys. 81 (5) (2010), https://doi.org/10.1103/PhysRevE.81.050903

[56] M.M. Müller, M. Deserno, J. Guven, Interface-mediated interactions between particles: a geometrical approach, Phys. Rev. E – Stat. Nonlinear Soft Matter Phys. 72 (6) (2005), https://doi.org/10.1103/PhysRevE.72.061407

[57] R.C. Van Lehn, A. Alexander-Katz, Membrane-embedded nanoparticles induce lipid rearrangements similar to those exhibited by biological membrane proteins, J. Phys. Chem. B 118 (44) (2014) 12586–12598, https://doi.org/10.1021/jp506239p

[58] Y. Li, X. Chen, N. Gu, Computational investigation of interaction between nanoparticles and membranes: hydrophobic/hydrophilic effect, J. Phys. Chem. B 112 (51) (2008) 16647–16653, https://doi.org/10.1021/jp8051906

[59] T.R. Weikl, Membrane-mediated cooperativity of proteins, Annu. Rev. Phys. Chem. 69 (1) (2018) 521–539, https://doi.org/10.1146/annurev-physchem-052516-050637

[60] J. Gao, R. Hou, L. Li, J. Hu, Membrane-mediated interactions between protein inclusions, Front. Mol. Biosci. 8 (2021).

[61] M. Karimi, J. Steinkühler, D. Roy, R. Dasgupta, R. Lipowsky, R. Dimova, Asymmetric ionic conditions generate large membrane curvatures, Nano Lett. 18 (12) (2018) 7816–7821, https://doi.org/10.1021/acs.nanolett.8b03584

[62] M. Goulian, R. Bruinsma, P. Pincus, Long-range forces in heterogeneous fluid membranes, EPL 22 (2) (1993) 145, https://doi.org/10.1209/0295-5075/22/2/012

[63] E. Lavagna, Z.P. Güven, D. Bochicchio, F. Olgiati, F. Stellacci, G. Rossi, Amphiphilic nanoparticles generate curvature in lipid membranes and shape liposome–liposome interfaces, Nanoscale 13 (40) (2021) 16879–16884, https://doi.org/10.1039/D1NR05067B

[64] E. Lavagna, J. Barnoud, G. Rossi, L. Monticelli, Size-dependent aggregation of hydrophobic nanoparticles in lipid membranes, Nanoscale 12 (17) (2020) 9452–9461, https://doi.org/10.1039/D0NR00868K

[65] D. Vasudevan, R.R. Gaddam, A. Trinchi, I. Cole, Core-shell quantum dots: properties and applications, J. Alloys Compd. 636 (2015) 395–404, https://doi.org/10.1016/j.jallcom.2015.02.102

[66] J. Cardellini, L. Caselli, E. Lavagna, S. Salassi, H. Amenitsch, M. Calamai, et al., Membrane phase drives the assembly of gold nanoparticles on biomimetic lipid bilayers, J. Phys. Chem. C 126 (9) (2022) 4483–4494, https://doi.org/10.1021/acs.jpcc.1c08914

[67] O. Bitton, S.N. Gupta, G. Haran, Quantum dot plasmonics: from weak to strong coupling, Nanophotonics 8 (4) (2019) 559–575, https://doi.org/10.1515/nanoph-2018-0218

[68] L. Caselli, A. Ridolfi, J. Cardellini, L. Sharpnack, L. Paolini, M. Brucale, et al., A plasmon-based nanoruler to probe the mechanical properties of synthetic and biogenic nanosized lipid vesicles, Nanoscale Horiz. 6 (7) (2021) 543–550, https://doi.org/10.1039/D1NH00012H

[69] C. Montis, L. Caselli, F. Valle, A. Zendrini, F. Carlà, R. Schweins, et al., Shedding light on membrane-templated clustering of gold nanoparticles, J. Colloid Interface Sci. 573 (2020) 204–214, https://doi.org/10.1016/j.jcis.2020.03.123

[70] F. Tian, T. Yue, Y. Li, X. Zhang, Computer simulation studies on the interactions between nanoparticles and cell membrane, Sci. China Chem. 57 (12) (2014) 1662–1671, https://doi.org/10.1007/s11426-014-5231-7
[71] M.R. Rasch, E. Rossinyol, J.L. Hueso, B.W. Goodfellow, J. Arbiol, B.A. Korgel, Hydrophobic gold nanoparticle self-assembly with phosphatidylcholine lipid: membrane-loaded and janus vesicles, Nano Lett. 10 (9) (2010) 3733–3739, https://doi.org/10.1021/nl102387n
[72] M. Daniel, J. Řezníčková, M. Handl, A. Iglič, V. Kralj-Iglič, Clustering and separation of hydrophobic nanoparticles in lipid bilayer explained by membrane mechanics, Sci. Rep. 8 (1) (2018) 10810, https://doi.org/10.1038/s41598-018-28965-y
[73] E. Rascol, J.-M. Devoisselle, J. Chopineau, The relevance of membrane models to understand nanoparticles–cell membrane interactions, Nanoscale 8 (9) (2016) 4780–4798, https://doi.org/10.1039/C5NR07954C
[74] G. Zuo, Q. Huang, G. Wei, R. Zhou, H. Fang, Plugging into proteins: poisoning protein function by a hydrophobic nanoparticle, ACS Nano 4 (12) (2010) 7508–7514, https://doi.org/10.1021/nn101762b
[75] O.K. Nag, M.E. Muroski, D.A. Hastman, B. Almeida, I.L. Medintz, A.L. Huston, et al., Nanoparticle-mediated visualization and control of cellular membrane potential: strategies, progress, and remaining issues, ACS Nano 14 (3) (2020) 2659–2677, https://doi.org/10.1021/acsnano.9b10163
[76] K. Park, S. Weiss, Design rules for membrane-embedded voltage-sensing nanoparticles, Biophys. J. 112 (4) (2017) 703–713, https://doi.org/10.1016/j.bpj.2016.12.047
[77] K. Park, Y. Kuo, V. Shvadchak, A. Ingargiola, X. Dai, L. Hsiung, et al., Membrane insertion of—and membrane potential sensing by—semiconductor voltage nanosensors: feasibility demonstration, Sci. Adv. 4 (1) (2018) e1601453, https://doi.org/10.1126/sciadv.1601453
[78] A. Torchi, F. Simonelli, R. Ferrando, G. Rossi, Local enhancement of lipid membrane permeability induced by irradiated gold nanoparticles, ACS Nano 11 (12) (2017) 12553–12561, https://doi.org/10.1021/acsnano.7b06690
[79] Y. Chen, Y. Gao, Y. Huang, Q. Jin, J. Ji, Inhibiting quorum sensing by active targeted pH-sensitive nanoparticles for enhanced antibiotic therapy of biofilm-associated bacterial infections, ACS Nano 17 (11) (2023) 10019–10032, https://doi.org/10.1021/acsnano.2c12151
[80] X. Ding, C. Yin, W. Zhang, Y. Sun, Z. Zhang, E. Yang, et al., Designing aptamer-gold nanoparticle-loaded pH-sensitive liposomes encapsulate morin for treating cancer, Nanoscale Res. Lett. 15 (1) (2020) 68, https://doi.org/10.1186/s11671-020-03297-x
[81] H. Jang, L.E. Pell, B.A. Korgel, D.S. English, Photoluminescence quenching of silicon nanoparticles in phospholipid vesicle bilayers, J. Photochem. Photobiol. A: Chem. 158 (2) (2003) 111–117, https://doi.org/10.1016/S1010-6030(03)00024-8

CHAPTER THREE

Exploring interactions between lipid membranes and nanoparticles through neutron and X-ray reflectometry techniques

Yuri Gerelli[a,b,*]
[a]Institute for Complex Systems, National Research Council, Piazzale Aldo Moro, Rome, Italy
[b]Department of Physics, Sapienza University of Rome, Piazzale Aldo Moro, Rome, Italy
*Corresponding author. e-mail address: Yuri.gerelli@roma1.infn.it

Contents

Abstract

Nanoparticle–membrane interactions play a pivotal role in various biological and biomedical processes, making their study essential for understanding nanoparticle behavior in complex biological systems. In this chapter, we explore the application of neutron and X-ray reflectometry (NR and XRR) techniques in investigating the structural dynamics and functional consequences of nanoparticle–membrane interactions, with a particular focus on soft and hard nanoparticles interacting with lipid membranes.

We begin by discussing the advantages of NR and XRR techniques, which provide valuable insights into the internal structure of lipid membranes, the spatial distribution of nanoparticles in the proximity or within solid supported lipid bilayers (SLBs).

Furthermore, we delve into the unique information obtained through NR experiments, where the contrast variation method allows for the quantification of lipid, water, and nanoparticle components. This enables the differentiation between

Advances in Biomembranes and Lipid Self-Assembly, Volume 38
ISSN 2451-9634, https://doi.org/10.1016/bs.abl.2023.07.001

structural modifications induced by nanoparticle incorporation and the formation of water-filled pores or membrane disruption.

Several case studies are presented, highlighting the use of NR and XRR techniques to discuss the influence of nanoparticle surface charge, shape, and composition on the mode of interaction, membrane integrity, and lipid organization. Additionally, the role of membrane complexity, including the presence of natural lipid and protein components, in shaping nanoparticle–membrane interactions is emphasized.

Overall, this chapter demonstrates the significance of neutron and X-ray reflectometry techniques in unraveling the intricate details of nanoparticle–membrane interactions. The insights gained from these studies contribute to the development of safe and effective nanoparticle-based therapeutics, as well as the understanding of fundamental biological processes at the nanoscale.

1. Introduction

Understanding the interaction between nanoparticles (NPs) and lipid membranes or bilayers is of paramount importance in various fields, including nanomedicine, drug delivery systems, and materials science.

These interactions are governed by a combination of physical, chemical, and biological factors, including electrostatic, hydrophobic, van der Waals, and steric interactions, as well as biological recognition. The relative contribution of these driving forces depends on the specific characteristics of the NPs and lipid membranes, such as their composition, size, surface charge, and surrounding environment.

Hard NPs, composed of metallic or inorganic cores, can be unmodified or coated to prevent aggregation in aqueous solutions. The presence or absence of specific coatings has a profound impact on the interaction between nanoparticles and lipid membranes, as it significantly alters their surface properties and potential functionalities, particularly in theragnostic applications [1–4].

Soft NPs encompass a wide array of structures including liposomes, micelles, and polymers and can be engineered to encapsulate, among others, biomolecules as drugs, protein, DNA or RNA [5–7]. On the contrary of the hard NPs, soft ones can be highly deformable, can fuse with biomembranes forming new nanostructures and can be made responsive to external stimuli or environmental conditions. In the context of their interaction with lipid membranes, soft NPs are commonly investigated for their potential as drug delivery systems. More recently, during the COVID-19 pandemic, by utilizing soft nanoparticles that mimic viral capsids, researchers have been able to study the behavior of viral proteins and their interactions with lipid membranes in a controlled and safe manner [8].

The complexity of NPs–membrane interactions arises from the combination of multiple driving forces, necessitating a comprehensive understanding to optimize NP-based systems for diverse applications. Different scenarios can occur when hard NPs interact with lipid membranes, including the reorganization of lipid components, integration within the membrane without compromising its integrity, or remaining in close proximity to the membrane without penetration. Soft NPs share these scenarios and also have the potential for fusion with the membrane. In all cases, any cargo carried by the NPs on their surface, within their coating or in their core (the latter for soft NPs only) can be released in the membranes.

To elucidate the result of NPs–membrane interactions at the molecular level, advanced characterization techniques are required. Neutron and X-ray reflectometry have emerged as powerful tools for investigating the structural features of this type of systems [9–16]. However, their routine application in this specific context has been limited thus far. This can be attributed to several factors, including the inherent complexity of the experimental setup, the specialized nature of the instrumentation required, and the challenges associated with data analysis. Recent advancements in instrumentation, sample preparation techniques, and data analysis algorithms have started to address these challenges, making neutron and X-ray reflectivity techniques increasingly accessible and applicable to the study of nanoparticle–lipid membrane interactions [17–23].

1.1 Neutron reflectometry: Principles and capabilities

Neutron reflectometry is a versatile experimental technique that utilizes the sensitivity of neutrons to the nuclear composition of a sample and their ability to penetrate deep into materials [24]. Neutrons are non-destructive probes capable of distinguishing between isotopes of the same element making NR particularly suitable for studying biological systems that utilize hydrogen and deuterium labeling to enhance the sensitivity of the neutrons to certain regions or molecules within the sample [25,26]. Within the field of biological sciences, planar lipid membranes, also known as supported lipid bilayers (SLBs), have been extensively investigated using NR due to their nanometric thickness and macroscopic lateral size.

In an NR experiment, a collimated neutron beam is directed at the sample, which is deposited on a flat surface or interface. As neutrons interact with the sample, they may undergo several scattering events;

among them, reflections are the events giving rise to the quantity that is measured, reflectivity, and they occur at each encountered interface. In NR, an interface is defined as a plane in space, perpendicular to the plane of incidence of the neutron beam [18,19,27], at which the refractive index, as seen by the probe, varies. In good approximation, the refractive index seen by a cold neutron is represented by the scattering length density (SLD or ρ), a quantity that depends on the nuclear composition of the sample and is sensitive to changes in nuclear and isotopic composition [28] Therefore, it is crucial to consider that two chemically diverse portions of the sample can possess similar or identical SLD values, making them indistinguishable to the neutron beam. On the other hand, isotopic labeling can provide a means to clearly distinguish regions with similar chemical compositions from each other.

In general, the SLD for N nuclei occupying a volume V can be calculated as

$$SLD = \frac{\sum_{j=1}^{N} b_j}{V}. \tag{1}$$

where b_j is the coherent scattering length of the jth nucleus. Values for naturally abundant nuclei and their isotopes can be found in literature [29]. Upon one or a series of reflections, the total number of neutrons that are reflected at an angle equal to that of the incident beam is determined by measuring the specularly reflected intensity using a suitable detector. The ratio between the number of reflected and incident neutrons gives rise to the reflectivity R. To obtain meaningful data, the incident angle (θ) of the neutron beam and/or the neutron wavelength (λ) are varied, and the reflectivity is determined for each configuration. Changing θ or λ results in probing different regions of the reciprocal space identified by the exchanged wave-vector $Q = \frac{4\pi \cdot \sin(\theta)}{\lambda}$. The collection of reflectivity values measured at different Q values is commonly indicated as $R(Q)$.

In a specular reflectometry experiment, only information regarding the sample's composition and structure along the perpendicular direction with respect to the surface/interface of reference can be obtained. Lateral "in-plane" information is averaged during the measurement. In the case of an SLB, this results in the identification of typically four internal regions that can be characterized during data analysis, as shown in Fig. 1. These regions correspond to the hydrophilic (head-groups) and hydrophobic (tails) regions of the bilayer. Generally, these regions have different SLD values, and when

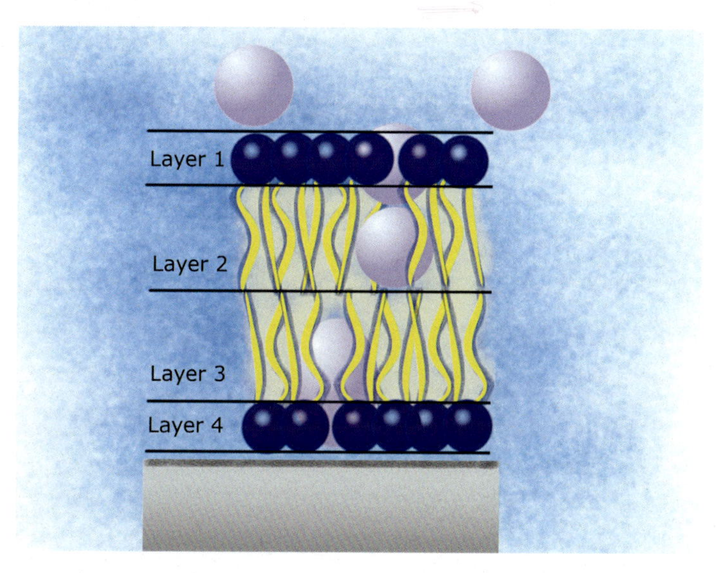

Fig. 1 Pictorial representation of an SLB on a solid substrate with nanoparticles. The diagram illustrates the different layers within the SLB. It is important to note that additional layers, including the substrate, the water gap between the substrate and the SLB, and the NPs on the top, are not depicted but should be considered for accurate modeling purposes.

a neutron crosses the boundary between two regions, a reflection event can occur. By selectively labeling isotopes (notably through partial deuteration), the number of internal layers can be varied. For each "visible" layer, the structural parameters that can be obtained from a modeling procedure include the thickness, composition (expressed in terms of average SLD), and interfacial roughness [30]. The most commonly used representation is the slab model [22], although alternative modeling approaches based on the reconstitution of the volume fraction profile of the membrane component have recently started to be used [31]. For more detailed information about modeling planar membranes for NR experiments, please refer to references [9,22].

In the context of nanoparticle–SLB interactions, NR can provide valuable information regarding the localization and distribution of nanoparticles within or in proximity to the membrane. Additionally, it offers insights into the structural modifications that nanoparticles might induce in the membrane. These modifications can include the reorganization of membrane components, pore formation, delivery of cargo molecules, and membrane disruption.

As already mentioned, a great advantage of the NR technique is the isotopic substitution. This can be achieved either by selective deuteration of specific regions or components within the sample but also by the replacement of H_2O with D_2O or H_2O:D_2O mixtures. In all cases, this procedure alters the difference in SLD, i.e. the contrast, between the regions within the sample and between the sample and the surrounding solvent. This method is commonly called contrast variation sample [25,26].

By acquiring reflectivity curves at multiple isotopic contrasts, it becomes possible to map the spatial distribution of nanoparticles along the vertical axis wither in the proximity, on the surface or within a lipid bilayer. Moreover, structural modifications induced by the presence of the NPs on the SLB can be investigated with an enhanced detail. Finally, the use of multiple isotopic contrasts aids in refining and validating the model, leading to a more accurate representation of the nanoparticle–membrane system and facilitating the extraction of valuable insights from the experimental results.

1.2 X-ray reflectometry: Principles and capabilities

The working principles of X-ray reflectometry (XRR) are very similar to those of NR [24]. The key difference between the two techniques lies in the nature of the radiation employed. XRR utilizes either a lab-based X-ray source or synchrotron radiation, which interact with electron clouds, providing information about the electron density distribution in the sample. Unlike neutron reflectometry, X-ray photons are insensitive to isotopic labeling. Consequently, in an XRR experiment, only one contrast condition can be used. This limitation has resulted in XRR being less commonly utilized for investigating biological matter. On the other hand, XRR offers enhanced performances in terms of both structural resolution (higher than that achievable using NR) and in sensitivity to metals, conditions particularly important when investigating metal-based NPs interacting with membranes.

For XRR, the probability of a specular reflection event is determined by the variation in electron density, ED. ED can be expressed as

$$ED = \frac{\sum_{j=1}^{N} e_j}{V} \tag{2}$$

where e_j is the number of electrons in the jth atom. Eq. (2) can be expressed in the same units of the SLD (Eq. (1)) if multiplied by the classical electron radius ($r_e = 2.8179\ fm$).

1.3 Complementary techniques and experimental considerations

Neutron and X-ray reflectometry are often combined with other techniques to obtain a more comprehensive understanding of the nanoparticle–membrane interactions. For instance, small-angle or dynamic light scattering can provide information about the size, shape, and aggregation behavior of nanoparticles in solution before their interaction with lipid membranes.

Additionally, techniques such as quartz-crystal microbalance with dissipation monitoring (QCM-D) are utilized to quantify the kinetics of adsorption or interaction, measure the net adsorbed mass, and assess changes in the viscoelastic properties of the membrane films. QCM-D not only aids in characterizing the interaction between nanoparticles and lipid membranes but also plays a crucial role in determining the protocols for sample preparation in reflectometry experiments. Given the similar experimental conditions, QCM-D assists in establishing the procedures required to prepare well-defined lipid membranes with controlled composition and effectively incorporate nanoparticles into the system. Ensuring the integrity of the samples is of utmost importance, and this involves careful attention to factors such as the use of suitable surfactants, solvent exchange techniques, and optimization of the lipid-to-nanoparticle ratio. These considerations are vital for achieving stable and reproducible lipid–nanoparticle systems.

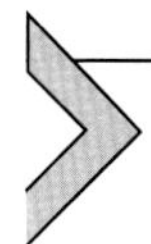

2. Nanoparticles and planar membranes: Adsorption, penetration and fusion processes seen by reflectometry techniques

Hard NPs, such as gold (Au), titanium (Ti), and superparamagnetic iron oxide (SPIONs), possess exceptional optical, electronic, and magnetic properties, making them versatile platforms for theragnostic applications [32]. Au NPs exhibit unique optical properties, including surface plasmon resonance, enabling precise control over their absorption and scattering properties in the visible and near-infrared regions. This feature makes them highly suitable for imaging techniques, such as photoacoustic imaging and surface-enhanced Raman spectroscopy, facilitating non-invasive visualization and detection of disease sites [33]. Ti-based NPs, with their biocompatibility and corrosion resistance, offer opportunities for biomedical applications, including biosensing, bioimaging, and implant coatings [34]. SPIONs possess unique magnetic properties, making them ideal for magnetic resonance imaging (MRI) and magnetic hyperthermia-based therapy. SPIONs can be manipulated by external magnetic fields, enabling their precise localization and control within biological systems [35].

Despite their immense potential, the interaction of hard NPs with lipid membranes can lead to cytotoxic effects, necessitating a comprehensive understanding of their biocompatibility and potential adverse outcomes [36]. When hard NPs encounter lipid membranes, several factors influence their cytotoxicity, including NP size, surface charge, composition, and surface functionalization. It is crucial to evaluate the influence of these factors on membrane integrity, cellular uptake, and intracellular responses. The cytotoxic effects can range from membrane damage and disruption of cellular processes to the generation of reactive oxygen species and inflammation. Thus, a balanced assessment of the potential therapeutic benefits and cytotoxicity is essential for the safe and effective utilization of hard NPs.

While several experimental techniques can be used to determine the net amount of material present on a surface and its evolution as a function of time or external stimuli, only NR and XRR can probe the internal structure of a membrane providing information on the NPs localization with respect to the different regions of the sample. Moreover, thanks to the contrast variation method, NR is particularly suitable to probe changes in composition of the sample and it is therefore capable of quantifying the amount of lipid, water molecules and nanoparticles. In this way structural changes, including those indicating membrane disruption such as formations of water-filled pores or extended lipid removal, due to NPs incorporation and adsorption can be characterized.

Gold NPs are probably the most used systems because they can be synthesized with precise control over their size and shape, they are highly stable, biocompatible and because they can be easily modified using various surface-chemistry protocols. One of the earliest investigations based on the use of NR on the potential cytotoxic effects of metallic NPs is the work published by Tatur et al. [37]. The authors performed a study on the interaction of gold nanoparticles (AuNPs, diameter 2 ± 0.5 nm) with different surface modifications (cationic and anionic surfaces) with zwitterionic lipid double bilayers composed of 1,2-distearoyl-*sn*-glycero-3-phosphocholine (DSPC). Structural information was obtained that provided insight into the fate of the AuNPs with regard to the integrity of the model cell membranes. In particular, they demonstrated that while cationic NPs could penetrate the membrane and remained embedded in the hydrophobic moiety of the floating bilayer, anionic ones remained on the bilayer surface causing bilayer dehydration, without any clear indication of cytotoxicity or membrane poration.

In a more recent work, Lolicato et al. [13] investigated the role of temperature and NPs surface charge on the interaction between cationic functionalized gold NPs (4 nm diameter) and SLB composed of zwitterionic (DSPC) and anionic (di-stearoyl-phosphatidylglycerol, DSPG) phospholipids by combining NR with molecular dynamics (MD) simulations. DSPG molecules were added to the zwitterionic components to mimic the surface of plasma membranes that are notably negatively charged. In the case of a pure DSPC SLB both MD simulations and NR experiments detected the penetration (irreversible) of the AuNPs within the lipid bilayer once the lipids were in the fluid phase, in agreement with the earlier results by Tatur. In this case, the structure of the membrane resulted perturbed but not destroyed; in particular, as hydration of the lipid headgroups resulted reduced, the authors hypothesized a change in molecular packing and order. In contrast, the AuNP adsorption resulted weak and impaired upon heating in the case of a partially charged DSPC–DSPG (3:1 by mol.) SLB. The presence of DSPG facilitated the aggregation of AuNPs, that resulted located on the SLB surface, and induced lipid crawling, causing membrane instability. Molecular dynamics simulations provided further insights into the interactions and revealed effects such as membrane puncturing, lipid concentration inhomogeneity and a potentially harmful AuNP aggregation effect for negative bilayers containing a large amount (25 mol% or higher) of negatively charged lipids [13].

A very similar study was conducted and published by Pfeiffer et al. [16] using XRR in combination with confocal microscopy, fluorescence correlation spectroscopy and MD simulations for the study of cationic AuNPs interacting with zwitterionic and anionic SLBs. Despite these general similarities, both lipid composition and NPs surface chemistry differed from the experiments by Tatur [37] and Lolicato [13]. In the present case, the SLB was composed of POPC (1-palmitoyl-2-oleoyl-*sn*-glycero-3-phosphocholine) and POPG (1-palmitoyl-2-oleoyl-*sn*-glycero-3-phospho-(1′-rac-glycerol)) in a 1:1 molar ratio, while AuNPs were functionalized with cationic molecules. The techniques chosen allowed the authors to probe different length scales, leading to the characterization of different stages of NP–SLB interaction. First, they determined that AuNPs adhesion to the SLB was induced by a strong electrostatic attraction and by an entropic contribution due to counterion condensation. Upon adhesion, the AuNPs caused an overall decrease of the SLB thickness, and an increase of its roughness as determined by XRR (top panel and inset in Fig. 2). The absence of clear increase in electron density (called SLD in the inset) in the

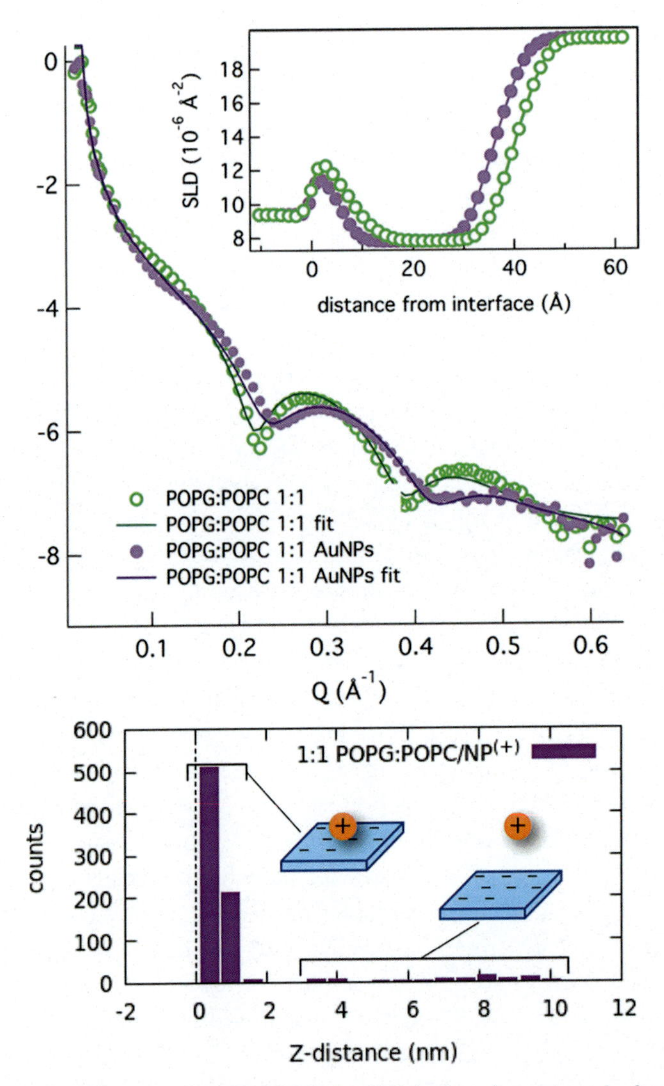

Fig. 2 (Top) XRR curves of POPG:POPC 1:1 SLB before (green) and after (purple) incubation with AuNPs; model curves (continuous lines) and corresponding electron density profiles (inset) are also reported. (Bottom) Occurrence distribution of the center-of-mass distance between the AuNP and the membrane in the *z*-direction calculated from MD simulations (the dashed vertical line at 0 represents the midplane *z*-quote). *Adapted with permission from T. Pfeiffer, Nanoparticles at biomimetic interfaces: combined experimental and simulation study on charged gold nanoparticles/lipid bilayer interfaces, J. Phys. Chem. Lett. 10 (2) (2019) 129–137. Copyright 2019 American Chemical Society [16].*

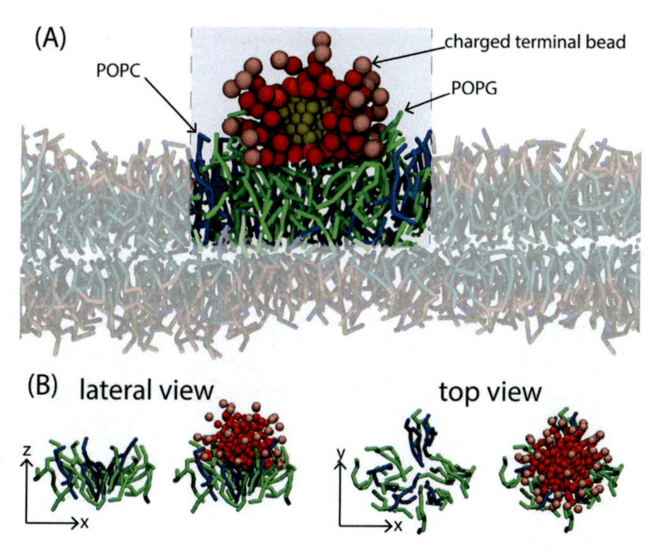

Fig. 3 (A) Lateral section of an AuNP interacting with a 1:1 POPG:POPC SLB determined by MD simulations. (B) Lateral and top views of the lipid selection reported in the upper panel. The AuNP is reported in red, while all other lipids and water solvent have been omitted for clarity. *Adapted with permission from T. Pfeiffer, Nanoparticles at biomimetic interfaces: combined experimental and smulation study on charged gold nanoparticles/lipid bilayer interfaces, J. Phys. Chem. Lett. 10 (2) (2019) 129–137. Copyright 2019 American Chemical Society [16].*

central region of the hydrophobic portion of the SLB ($z \sim 23$ Å in the inset of Fig. 2) suggests that cationic AuNPs do not penetrate the SLB. On the other hand, the small difference in electron density between water and the lipid molecules does not allow to determine if pores, filled with water, are formed upon interaction. MD simulations confirm this scenario, locating cationic AuNPs on the top of the SLB, as shown in the lower panel of Fig. 2 and in Fig. 3. As determined from MD simulations, an important role for establishing the interaction is given by counterion release and condensation. Indeed, MD simulations provided clear evidence that the features observed by XRR originated, at a molecular level, by the penetration of the positively charged ligand present on the AuNP surface into the headgroup region of the SLB. This causes a modification of the packing and orientation of lipid molecules inducing local reorganization of the POPC and POPG molecules as visible in Fig. 3.

The results reported in literature for AuNPs can be easily translated to NP having a different core composition but similar surface chemistry as this is the parameter crucial for establishing and driving the interaction. A role equally crucial is played by the SLB composition and in this respect the variability of the studies reported in literature is enormous given the huge number of different lipid molecules available and present in nature. Luchini et al. investigated the interaction of neutral and charged Super Paramagnetic Iron Oxide Nanoparticles (SPIONs) with SLB increasingly complex in composition, i.e. composed of different molar ratios of POPC and POPG and with and without cholesterol [10]. SPIONs of 6 nm diameter were stabilized either with a cationic surfactant (cetyltrimethyl ammonium bromide, CTAB) or a zwitterionic phospholipid (1-stearoyl-2-hydroxy-3glycero-*sn*-phosphocholine, 18LPC). The results reported in [10] concerning both the cationic and zwitterionic SPIONs are in agreement with the observation by Pfeiffer [16] on cationic AuNPs and negatively charged SLB; also SPIONs are found on the "top" of the SLB i.e., on the surface of the system directly accessible to the aqueous phase with no clear signature of membrane disruption. Despite being a common feature for all investigated samples, the authors could establish a link between the amount of SPIONs lying on the SLB surface that was larger for samples containing a larger amount of cholesterol. Differences in the absorbed amount were also induced, for the same SLB composition, by the SPION functionalization, with 18LPC-SPIONs irreversibly bound to the SLB "top" surface giving rise to a highly hydrated NP layer. Based on the irreversibility of the adsorption process the authors hypothesized that zwitterionic SPIONs might accumulated continuously on the cell surface making them potential cytotoxic compounds.

While cytotoxicity must be avoided or minimized to the fullest extent in many applications such as drug delivery and diagnostics, membrane disruption serves as an effective killing mechanism in certain therapeutic approaches, as for example against bacterial or cancer cells.

In this context, Malekkhaiat Häffner et al. [15] conducted an investigation on the utilization of titanium-oxide (TiO_2) NPs in conjunction with UV illumination to induce oxidative degradation of bacterial lipids. The experiments were carried out under acidic pH conditions (pH 3.4), where the NPs exhibited a positive charge and remained stable in solution, with an average size of 8 ± 1 nm. It is noteworthy that, unlike previous cases reported, these TiO_2 NPs were devoid of any ligands or molecules on their surface. Among the various techniques employed, neutron reflectivity was used to investigate the structural changes induced by TiO_2 NPs alone or in combination with UV illumination on negatively charged SLBs composed of POPC, POPG, and a polyunsaturated lipid, palmitoylarachidonoyl phosphocholine (PAPC). PAPC was chosen due to its prevalence as one of the most abundant polyunsaturated phospholipids in nature. By varying the PAPC content in the SLB, the authors were able to elucidate the effects of polyunsaturation on the oxidative destabilization of lipid membranes. NR not only facilitated the determination of structural changes resulting from NP binding and UV illumination but also enabled the assessment of the time-dependent evolution of photooxidative effects. Fig. 4 presents a representative summary of this intricate investigation. Panel A displays time-resolved NR data for an SLB composed of POPC, POPG, and PAPC (2:1:1 molar ratio). The initial reflectivity (gray symbols) exhibits a higher intensity, indicating the pristine state of the SLB in the absence of both NPs and illumination. Upon the addition of TiO_2 NPs (black symbols), the reflectivity decreases, suggesting a reduction in contrast due to the partial removal of lipids and the formation of water-accessible pores. Subsequent exposure to UV light (approx. 254 nm wavelength) leads to further continuous decreases in reflectivity and to changes in the shape of the $R(Q)$ curves suggesting structural modifications or molecular rearrangement in the SLB. All these changes are rationalized by the modeling of the NR data. Fig. 4B shows the SLD profiles corresponding to the model $R(Q)$ curves shown in Fig. 4A (solid black lines). They indicate that the SLD of the bilayer increases continuously towards the value of the solvent used for this experiment, which is heavy water (D_2O), in agreement with lipid removal and the formation of water-accessible pores.

Another prominent feature that emerges from the analysis is the formation of a highly hydrated layer on top of the SLB, with a thickness comparable to that of the TiO_2 NPs. By its SLD value, this layer is composed of NPs and lipid molecules. The amount of material present on the top of the SLB increases with time but reaches saturation after 40 min of UV exposure as indicated by the overlapping SLD profiles for $70 < z < 100$ nm. Analyzing the structural parameters associated with these models, the authors identified a monotonous increase, as a function of exposure time, in the area per molecule (panel C) and in lipid removal (panel D). By conducting a similar analysis on SLBs with different compositions, the authors concluded that the photooxidative effect is enhanced in the presence of NPs and anionic POPG lipids, while it is reduced in the presence of cholesterol (data not shown here). Combining these findings with the results from complementary techniques, the authors were able to elucidate the intricate interplay between the presence of nanoparticles, UV light illumination, and membrane composition in photooxidative processes, as illustrated in the provided cartoon in Fig. 5. Specifically, the authors demonstrated that NP binding alone has a minor destabilizing effect on lipid bilayers of any

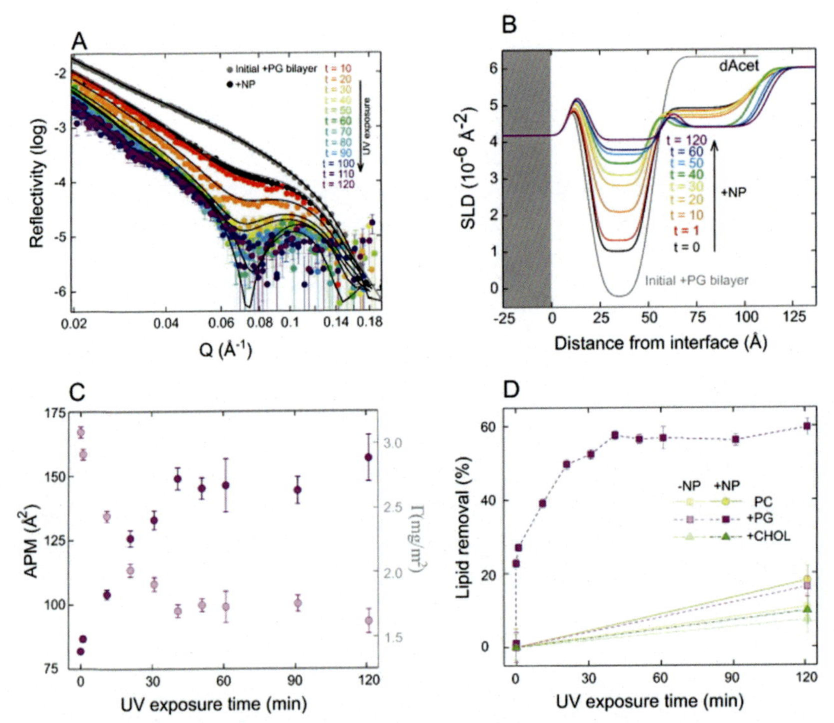

Fig. 4 (A) Time-resolved NR data (symbols) and model curves (solid lined) for a 2:1:1 POPC:POPG:PAPC SLB in the presence of 100 ppm TiO_2 NPs during 2 h of in-situ UV illumination. (B) Evolution of the SLD profiles corresponding to the model curves reported in panel A. (C) Evolution of the area per molecule (AMP, left axis) and surface coverage (Γ, right axis) during the experiment. (D) Extent of lipid removal as a function of the exposure time for the three bilayer compositions (POPC + PAPC, POPC + PAPC + POPG, and POPC + PAPC + POPG + CHOL), with (+NP) or without (−NP) TiO_2 NPs. *Readapted from J.S. Malekkhaiat Häffner, E. Parra-Ortiz, M.W.A. Skoda, T. Saerbeck, K.L. Browning, M. Malmsten, Composition effects on photooxidative membrane destabilization by TiO_2 nanoparticles, Journal of Colloid and Interface Science, 584, 19–33, Copyright (2021), with permission from Elsevier [15].*

composition. Under UV illumination, TiO_2 NPs induce various changes depending on the membrane composition, including lipid removal, increased hydration, gradual bilayer thinning, lateral phase separation, and, under certain conditions, solubilization into aggregates that can accumulate on the top of the SLB or be solubilized in solution.

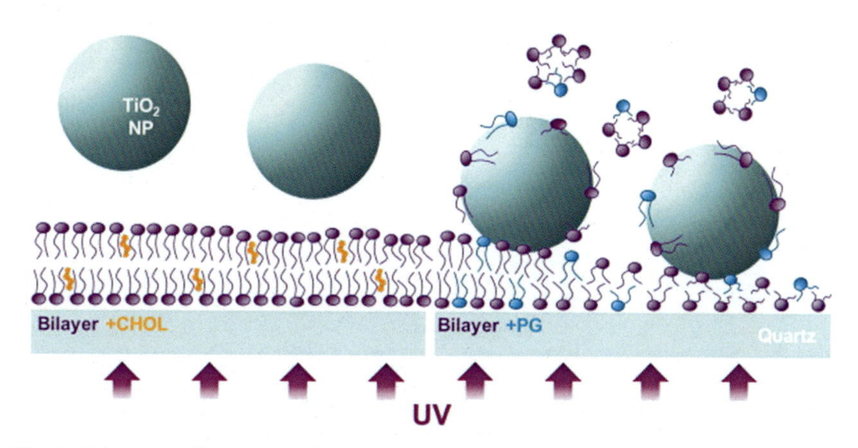

Fig. 5 Schematic illustration of structural changes induced by UV-activated TiO_2 NPs, including lipid removal, increases in hydration, gradual bilayer thinning, lateral phase separation, and formation of NP–lipid aggregates. *Readapted from S. Malekkhaiat Häffner, E. Parra-Ortiz, M.W.A. Skoda, T. Saerbeck, K.L. Browning, M. Malmsten, Composition effects on photooxidative membrane destabilization by TiO_2 nanoparticles, Journal of Colloid and Interface Science, 584, 19–33, Copyright (2021), with permission from Elsevier [15].*

3. Effect of membrane and nanoparticle complexity

As already introduced in the previous section, the interaction between nanoparticles and lipid bilayers is influenced by the presence of different lipid species and their diverse physical properties, such as chain length, saturation, and headgroup chemistry. These variations in the membrane composition give rise to distinct packing arrangements, lateral organization, and surface charge, which in turn can modulate the adsorption, insertion, or permeation of nanoparticles into the lipid bilayer. Furthermore, the complexity of the membrane is further heightened by the presence of membrane proteins and other biomolecules, including cholesterol and glycans. These additional components introduce an extra layer of intricacy to the interactions between nanoparticles and the membrane. Proteins embedded within the lipid bilayer or present on the membrane surface can serve as mediators or receptors for nanoparticle binding, leading to specific recognition and enhanced interactions. Therefore, gaining a comprehensive understanding of the influence of membrane complexity is crucial for elucidating the behavior of nanoparticles in biological systems. It also plays a vital role in the rational design of nanoparticles for various applications, ranging from drug delivery and biosensing to nanomedicine.

A recent advancement in this field involves the utilization of natural and biogenic SLBs, which exhibit a diverse range of components depending on their extraction from various natural sources. An exemplary study by Montis et al. [38] compared the effect of SPIONs on a biogenic SLB, formed using nanosized extracellular vesicles [39], with that on a synthetic zwitterionic SLB composed of POPC. The authors employed X-ray reflectivity in conjunction with other surface-sensitive techniques such as quartz crystal microbalance with dissipation monitoring (QCM-D), atomic force microscopy (AFM), and confocal microscopy.

By synthesizing the collective experimental findings, the authors discovered that the adhesion of SPIONs to the SLB is significantly enhanced when the latter is composed of biogenic lipids. Consistent with the findings reported in [10], SPIONs were observed to accumulate on the surface of the SLB; the surface density of the SPIONs layer resulted very low in the case of the zwitterionic bilayer. Furthermore, in this work, minimal structural deformation of the SLB was observed, suggesting that the interaction primarily occurs at the interface between the nanoparticles and the SLB surfaces. However, due to the inherent complexity of biogenic SLBs and the formation of a dense layer of SPIONs on top of the SLB, the structural modifications occurring within the biogenic SLB could not be resolved at the angstrom level. At the time of publication, modeling approaches were still limited, hindering a detailed understanding of these structural changes. Recent advancements in this field will be described in the next section.

Another relatively unexplored aspect in the field of nanoparticle–membrane interactions is the role of symmetry and shape anisotropy of both nanomaterials and lipid membranes. In a recent study, Caselli et al. [40] have investigated how gold nanospheres (AuNSs, 3.4 nm diameter) and gold nanorods (AuNRs, 2.6 diameter and 19.2 nm length) with similar size and surface coating interact with lipid films of different internal structure and morphology. Specifically, they examined a lamellar phase, a multi-supported lipid bilayer (multi-SLB) with 2D symmetry, and a cubic phase with 3D symmetry.

The use of lipid structures that are not made of a single lipid bilayer has a biological relevance. In fact, while supported lipid bilayers are commonly used as models for biological membranes, in certain cases such as during cell trafficking phenomena or under pathological conditions, membranes can undergo topological modifications, including a transition from a lamellar to a cubic bicontinuous phase. The transient nature of non-lamellar biological membranes presents significant challenges for their investigation in natural systems. By utilizing neutron reflectivity and complementary techniques, Caselli and colleagues determined that nanorods generally exhibit a stronger interaction with the lipid film compared to nanospheres. Furthermore, both types of nanoparticles induced larger structural modifications in the lamellar film, while the cubic phase was perturbed to a lesser extent.

Fig. 6 displays the NR data for the lamellar (A) and cubic (B) films in the absence (red) and presence (green) of AuNSs. Prior to the addition of nanoparticles, the samples exhibited the expected reflectivity curve respectively for a lamellar multilayer and a cubic phase, evident from the appearance of Bragg's peaks as indicated in the figure. Upon the addition of spherical nanoparticles, the overall symmetry of the structure was largely maintained, with the positions of the peaks remaining unaltered. However, the intensity of the Bragg's peaks in the lamellar phase decreased, indicating a partial disruption of the lamellar arrangement or a decrease in the number of repeating units in the sample. The effects of nanorods on the structure of both lamellar and cubic films were more pronounced and dramatic (data not reported, please refer to Fig. 4 in reference [40]). In this case, the initially ordered structures were destroyed, as evidenced by the disappearance of the Bragg's peaks. Moreover, the residual reflectivity resembled that of a single supported lipid bilayer on a silicon substrate. These results indicate that the drastic effect of nanorods is not dependent on the structure of the lipid film but rather on the shape anisotropy of the interacting particles.

Using confocal microscopy, the authors further investigated the structure of the samples at the micron-scale. Both nanospheres and nanorods induced structural and morphological changes in the lamellar and cubic

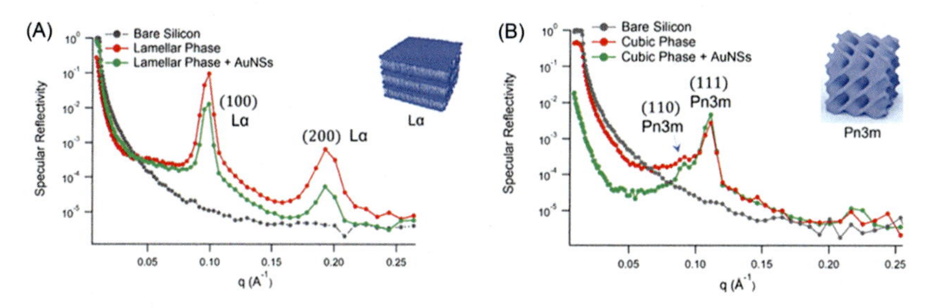

Fig. 6 NR data collected for the lamellar (A) and cubic (B) phases described in ref. [40] before (red data) and after (green data) addition of spherical AuNPs. Data of the bare silicon substrate are also shown for comparison (gray symbols). *Used with permission of Royal Society of Chemistry, from L. Caselli, et al., Interaction of nanoparticles with lipid films: the role of symmetry and shape anisotropy, 24 (2022); permission conveyed through Copyright Clearance Center, Inc.*

lipid phases. To summarize the experimental findings, the interaction resulted driven by both the asymmetry of the nanostructure of the lipid film and the shape anisotropy of the nanoparticles. The greater the asymmetry of the system, the stronger the interaction and the induced structural modifications.

Non-lamellar lipid phases serve not only as target membranes that mimic non-transient states representative of cellular processes but also as soft nanoparticles that can interact with single bilayered cell membranes.

Cubosomes are nanostructured particles characterized by a bicontinuous cubic liquid crystalline phase widely used for pharmaceutical and cosmetics applications [41]. Composed of amphiphilic components such as lipids and surfactants, often stabilized by polymers, cubosomes play a crucial role in various formulations. Investigating the interaction of cubosomes with lipid membranes is of fundamental importance not only for optimizing drug encapsulation, release kinetics, and stability within the membrane environment but also for assessing their potential toxicity or side effects on cell membranes to ensure product safety.

In this context, Shen and coworkers [11] investigated, by means of QCM-D and NR, the interaction of phytantriol-based cubosomes with a POPC SLB. In order to better distinguish between the molecular components of cubosomes and bilayer, the authors utilized partially deuterated POPC lipids (namely d_{31}-POPC). Cubosomes were added to a solution in the presence of a POPC SLB already formed on a solid substrate made of silicon oxide. It is worth mentioning that the authors conducted tests to investigate the interaction of these cubosomes with the bare silicon substrate, and the results demonstrated no adsorption. In the presence of a zwitterionic POPC SLB, both NR and QCM-D indicated the accumulation of cubosomes on the top of the SLB suggesting the establishment of an attractive interaction. NR experiments also provided valuable information on the changes induced in the structure and properties of the SLB as a result of the interaction with cubosomes as evidenced by the data presented in Fig. 7. NR measurements conducted in heavy water revealed that within 2 h of adding the cubosomes, the measured reflectivity significantly differed from that of the d_{31}-POPC SLB. The observed increase in reflectivity, considering the SLB's partial deuteration, indicated an increase in hydrogenated material at the interface. Additionally, the emergence of a Bragg peak ($Q \approx 0.14\,\text{Å}^{-1}$) indicated the presence of a well-defined periodic structure within the interfacial film. The observed periodicity (44.8 Å) aligned with the D-spacing previously measured by the

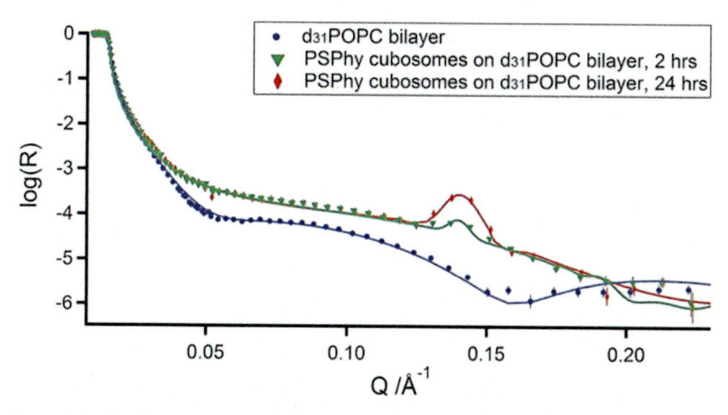

Fig. 7 NR data (symbols) and model curves (lines) for a d31-POPC bilayer in D_2O before (blue) and after incubation of cubosomes for 2 h (green) and 24 h (red). *Used with permission of Royal Society of Chemistry, from H-H. Shen, et al., The interaction of cubosomes with supported phospholipid bilayers using neutron reflectometry and QCM-D, 7 (2011); permission conveyed through Copyright Clearance Center, Inc.*

authors for the same cubosomes in solution [42]. After 24 h, the overall shape of the reflectivity remained unchanged, but the intensity of the Bragg peak increased, indicating a higher amount of cubosomes at the interface. Through a detailed analysis leveraging the deuteration of the SLB, the authors also identified surfactant–lipid exchange and an overall thinning of the SLB, both resulting from the penetration of the cubosomes into the bilayer.

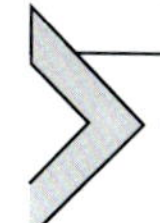

4. Towards more complex modeling approaches: Determination of the interaction distance between NPs and a membrane

Most of the examples of neutron reflectivity (NR) and X-ray reflectivity (XRR) presented thus far have employed simple modeling procedures. In these procedures, each discernible layer in the sample is represented as a slab with defined thickness, composition (expressed in terms of scattering length density or electron density), and interfacial roughness. This approach, known as the slab model, is well-suited for characterizing highly homogeneous samples with ordered planar structures. However, when applied to inherently non-homogeneous layers, such as layers containing nanoparticles (NPs), the insights obtained using this class of models are often limited.

For polymer layers at various interfaces, the scattering length density (SLD) or electron density (ED) profiles can be accurately described using numerical or analytical expressions. These profiles allow for precise characterization of the structural properties of the polymer layers and serve as the basis for calculating reflectivity curves [43,44]. However, when it comes to NP layers, the utilization of numerical or analytical expressions instead of the slab model is relatively limited in the published literature. This has limited the number of experiments on NP samples performed using reflectometry techniques and restricted the information that could be derived from the available experimental data.

In their recent paper, Armanious et al. [12] demonstrated the application of NR to investigate samples composed of biological nanoparticles interacting with planar SLBs. The study focused on the integrated structural characterization of hard silica (SiO_2) nanoparticles, soft nanoparticles (liposomes), and mixed nanoparticles (SiO_2 core coated with a lipid shell) in the presence of a POPC SLB. They utilized the slab model to describe the supporting surface and the SLB. To account for the contribution of the nanoparticles, the authors first calculated the projection of the nanoparticle volume in one dimension. This information was then used to generate discrete SLD profiles, corresponding to the chosen contrast conditions. Finally, the total SLD profile of the system was obtained by merging the contributions calculated with the slab model to that of the NP layer. Then, the total SLD profiles were converted into reflectivity curves, which were validated against experimental data. It should be noted that, in all investigated samples, the penetration of NPs into the underlying SLB was intentionally avoided to validate the method in the determination of the separation distance between the NPs and the membrane. The results of this approach, along with the derived information, are summarized in Fig. 8, which showcases the structural characterization of a POPC bilayer exposed to a solution containing silica nanoparticles.

It is important to acknowledge that due to the extensive number of structural parameters involved, certain parameters were fixed or constrained based on values obtained with higher accuracy using alternative techniques. For instance, the NP size was determined through microscopy techniques and dynamic light scattering, the material SLD fixed to known values, while the structure of the silicon substrate and SLB was determined by NR prior to the introduction of the NPs. Nonetheless, the developed model can still be employed even when these parameters are unknown. This flexibility allows for the application of the model in scenarios where precise values for these parameters may not be available, enabling the exploration of nanoparticle–membrane interactions under a broader range of conditions.

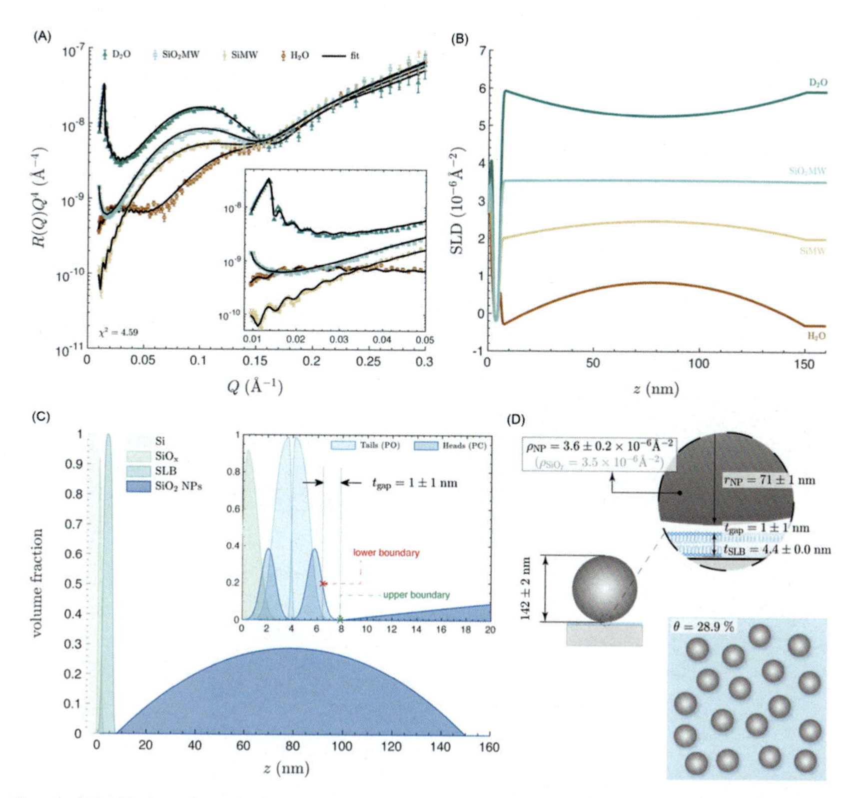

Fig. 8 (A) NR data (symbols) measured in four different contrast conditions for a POPC SLB after incubation of silica NPs. The black lines represent the theoretical curves corresponding to the volume fraction profile shown in (C). (B) SLD profiles determined from the volume fraction profile shown in (C) and used to calculate the model curves shown in (A). (C) Mono dimensional volume fraction profile. The inset provides details regarding the layer description of the underlying SLB. (D) Summary of all derived structural information. *Reproduced with permission from Armanious et al., J. Am. Chem. Soc. 144 (45) (2022) 20726 [12].*

As clearly visible in Fig. 8, the curvature of the NPS is reflected in the shape of their contribution to the SLD profiles that cannot be easily described using a finite (and low) number of slabs. Employing slabs with significant interfacial roughness or using a large number of extremely thin slabs (~1–2 Å thickness) could approximate the one-dimensional volume fraction profile of the nanoparticles and yield a reasonable model curve. However, the resulting structural information obtained remains qualitatively descriptive rather than highly accurate.

The described approach is not restricted solely to hard nanoparticles but has also been effectively employed to characterize liposomes and lipid-coated nanoparticles interacting at a nanometric distance with planar SLBs. In broader terms, this methodology can be further expanded to encompass nano-objects of diverse shapes and sizes, as long as their one-dimensional volume fraction profiles can be appropriately parameterized and numerically computed. This flexibility allows for the application of the approach to a wide range of nanostructures, enabling the investigation and characterization of their interactions with lipid membranes in various experimental systems.

5. Conclusions

This chapter has presented a comprehensive review of the utilization of neutron and X-ray reflectometry techniques for investigating the structural aspects of nanoparticle interactions with increasingly complex lipid membrane models. The case studies discussed herein have underscored the pivotal role played by various factors, including nanoparticle size, shape, surface properties, composition, as well as lipid membrane composition and phase behavior, in influencing the adsorption, insertion, and permeation of nanoparticles within lipid membranes. Through the refinement of our understanding of these interactions, neutron reflectivity (NR) and X-ray reflectivity (XRR) have made significant contributions to the rational design and optimization of nanoparticle-based systems across diverse fields such as drug delivery, biosensing, and nanomedicine. Reflectometry techniques have also proven to be ideal for precisely localizing nanoparticles with respect to supported lipid membranes. These techniques offer valuable information about the proximity of nanoparticles to lipid membranes or their insertion within the lipid layers. By analyzing the reflectivity curves and employing advanced modeling approaches, researchers have been able to accurately determine the separation distance and spatial arrangement of nanoparticles with respect to the lipid membranes.

By integrating cutting-edge modeling approaches with state-of-the-art instrumentation, future experiments in the field of nanoparticle–membrane interactions are expected to yield even more remarkable insights.

References

[1] M.P. Monopoli, C. Åberg, A. Salvati, K.A. Dawson, Biomolecular coronas provide the biological identity of nanosized materials, Nat. Nanotechnol. 7 (12) (2012) 779–786.

[2] M. Lundqvist, J. Stigler, G. Elia, I. Lynch, T. Cedervall, K.A. Dawson, Nanoparticle size and surface properties determine the protein corona with possible implications for biological impacts, Proc. Natl. Acad. Sci. U. S. A. 105 (2008) 14265–14270.

[3] A.K. Gupta, M. Gupta, Synthesis and surface engineering of iron oxide nanoparticles for biomedical applications, Biomaterials 26 (2005) 3995–4021.

[4] S. Laurent, D. Forge, M. Port, A. Roch, C. Robic, L. Vander Elst, et al., Magnetic iron oxide nanoparticles: synthesis, stabilization, vectorization, physicochemical characterizations and biological applications, Chem. Rev. 108 (2008) 2064–2110.

[5] S. Nayak, L. Andrew Lyon, Soft nanotechnology with soft nanoparticles, Angew. Chem. - Int. Ed. 44 (2005) 7686–7708.

[6] C. Horejs, From lipids to lipid nanoparticles to mRNA vaccines, Nat. Rev. Mater. 6 (12) (2021) 1075–1076.

[7] J. Barauskas, M. Johnsson, F. Tiberg, Self-assembled lipid superstructures: beyond vesicles and liposomes, Nano Lett. 5 (2005) 1615–1619.

[8] M.A. Heinrich, B. Martina, J. Prakash, Nanomedicine strategies to target coronavirus, Nano Today 35 (2020) 100961.

[9] Y. Gerelli, Applications of neutron reflectometry in biology, EPJ Web Conf. 236 (2020) 04002.

[10] A. Luchini, Y. Gerelli, G. Fragneto, T. Nylander, G.K. Pálsson, M.-S.S. Appavou, et al., Neutron reflectometry reveals the interaction between functionalized SPIONs and the surface of lipid bilayers, Colloids Surf. B Biointerfaces 151 (2017) 76–87.

[11] H.H. Shen, P.G. Hartley, M. James, A. Nelson, H. Defendi, K.M. McLean, The interaction of cubosomes with supported phospholipid bilayers using neutron reflectometry and QCM-D, Soft Matter 7 (2011) 8041–8049.

[12] A. Armanious, Y. Gerelli, S. Micciulla, H.P. Pace, R.J.L. Welbourn, M. Sjöberg, et al., Probing the separation distance between biological nanoparticles and cell membrane mimics using neutron reflectometry with sub-nanometer accuracy, J. Am. Chem. Soc. 144 (2022).

[13] F. Lolicato, L. Joly, H. Martinez-Seara, G. Fragneto, E. Scoppola, F. Baldelli Bombelli, et al., The role of temperature and lipid charge on intake/uptake of cationic gold nanoparticles into lipid bilayers, Small 15 (2019) e1805046.

[14] P. Vandoolaeghe, A.R. Rennie, R.A. Campbell, T. Nylander, Neutron reflectivity studies of the interaction of cubic-phase nanoparticles with phospholipid bilayers of different coverage, Langmuir 25 (2009) 4009–4020.

[15] S. Malekkhaiat Häffner, E. Parra-Ortiz, M.W.A. Skoda, T. Saerbeck, K.L. Browning, M. Malmsten, Composition effects on photooxidative membrane destabilization by TiO2 nanoparticles, J. Colloid Interface Sci. 584 (2021) 19–33.

[16] T. Pfeiffer, A. De Nicola, C. Montis, F. Carlà, N.F.A. Van Der Vegt, D. Berti, et al., Nanoparticles at biomimetic interfaces: combined experimental and simulation study on charged gold nanoparticleslipid bilayer interfaces, J. Phys. Chem. Lett. 10 (2019).

[17] A. Junghans, E.B. Watkins, R.D. Barker, S. Singh, M.J. Waltman, H.L. Smith, et al., Analysis of biosurfaces by neutron reflectometry: from simple to complex interfaces, Biointerphases 10 (2015) 019014.

[18] T. Saerbeck, R. Cubitt, A. Wildes, G. Manzin, K.H. Andersen, P. Gutfreund, Recent upgrades of the neutron reflectometer D17 at ILL, J. Appl. Crystallogr. 51 (2018) 249–256.

[19] R.A. Campbell, H.P. Wacklin, I. Sutton, R. Cubitt, G. Fragneto, FIGARO: the new horizontal neutron reflectometer at the ILL, Eur. Phys. J. Plus 126 (2011) 107.

[20] J. Webster, S. Holt, R. Dalgliesh, INTER the chemical interfaces reflectometer on target station 2 at ISIS, Phys. B Condens. Matter 385–386 (2006) 1164–1166.
[21] H.P. Wacklin, Neutron reflection from supported lipid membranes, Curr. Opin. Colloid Interface Sci. 15 (2010) 445–454.
[22] Y. Gerelli, Aurore: new software for neutron reflectivity data analysis, J. Appl. Crystallogr. 49 (2016) 330–339.
[23] A.R.J. Nelson, S.W. Prescott, Refnx: neutron and X-ray reflectometry analysis in python, J. Appl. Crystallogr. 52 (2019) 193–200.
[24] J. Daillant, A. Gibaud, X-ray and Neutron Reflectivity, Springer Berlin Heidelberg, Berlin, Heidelberg, 2009.
[25] T.L. Crowley, E.M. Lee, E.A. Simister, R.K. Thomas, The use of contrast variation in the specular reflection of neutrons from interfaces, Phys. B: Phys. Condens. Matter 173 (1991) 143–156.
[26] A. Nelson, Co-refinement of multiple-contrast neutron/X-ray reflectivity data using MOTOFIT, J. Appl. Crystallogr. 39 (2006) 273–276.
[27] F. Cousin, G. Fadda, An introduction to neutron reflectometry, EPJ Web Conf. 236 (2020) 04001.
[28] V.F. Sears, Fundamental aspects of neutron optics, Phys. Rep. 82 (1982) 1–29.
[29] V.F. Sears, Neutron scattering lengths and cross sections, Neutron News 3 (1992) 26–37.
[30] Y. Gerelli, Phase transitions in a single supported phospholipid bilayer: Real-time determination by neutron reflectometry, Phys. Rev. Lett. 122 (2019) 248101.
[31] M. Belička, Y. Gerelli, N. Kučerka, G. Fragneto, The component group structure of DPPC bilayers obtained by specular neutron reflectometry, Soft Matter 11 (2015) 6275–6283.
[32] N. Shreyash, M. Sonker, S. Bajpai, S.K. Tiwary, Review of the mechanism of nanocarriers and technological developments in the field of nanoparticles for applications in cancer theragnostics, ACS Appl. Bio Mater. 4 (2021) 2307–2334.
[33] D.A. Giljohann, D.S. Seferos, W.L. Daniel, M.D. Massich, P.C. Patel, C.A. Mirkin, Gold nanoparticles for biology and medicine, Angew. Chem. Int. Ed. 49 (2010) 3280–3294.
[34] S. Çeşmeli, C. Biray Avci, Application of titanium dioxide (TiO_2) nanoparticles in cancer therapies, J Drug Target 27 (2018) 762–766, https://doi.org/10.1080/1061186X.2018.1527338
[35] A.V. Samrot, C.S. Sahithya, J. Selvarani A, S.K. Purayil, P. Ponnaiah, A review on synthesis, characterization and potential biological applications of superparamagnetic iron oxide nanoparticles, Curr. Res. Green Sustain. Chem. 4 (2021) 100042.
[36] M. Nikzamir, A. Akbarzadeh, Y. Panahi, An overview on nanoparticles used in biomedicine and their cytotoxicity, J. Drug Deliv. Sci. Technol. 61 (2021) 102316.
[37] S. Tatur, M. MacCarini, R. Barker, A. Nelson, G. Fragneto, Effect of functionalized gold nanoparticles on floating lipid bilayers, Langmuir 29 (2013).
[38] C. Montis, A. Salvatore, F. Valle, L. Paolini, F. Carlà, P. Bergese, et al., Biogenic supported lipid bilayers as a tool to investigate nano-bio interfaces, J. Colloid Interface Sci. 570 (2020) 340–349.
[39] C. Montis, S. Busatto, F. Valle, A. Zendrini, A. Salvatore, Y. Gerelli, et al., Biogenic supported lipid bilayers from nanosized extracellular vesicles, Adv. Biosyst. 2 (2018) 1700200.
[40] L. Caselli, A. Ridolfi, G. Mangiapia, P. Maltoni, J.F. Moulin, D. Berti, et al., Interaction of nanoparticles with lipid films: the role of symmetry and shape anisotropy, Phys. Chem. Chem. Phys. 24 (2022) 2762–2776.
[41] S. Shetty, S. Shetty, Cubosome-based cosmeceuticals: A breakthrough in skincare, Drug. Discov. Today 28 (2023) 103623.

[42] H.H. Shen, J.G. Crowston, F. Huber, S. Saubern, K.M. McLean, P.G. Hartley, The influence of dipalmitoyl phosphatidylserine on phase behaviour of and cellular response to lyotropic liquid crystalline dispersions, Biomaterials 31 (2010) 9473–9481.
[43] E. Schneck, I. Berts, A. Halperin, J. Daillant, G. Fragneto, Neutron reflectometry from poly (ethylene-glycol) brushes binding anti-PEG antibodies: Evidence of ternary adsorption, Biomaterials 46 (2015) 95–104.
[44] E. Schneck, A. Schollier, A. Halperin, M. Moulin, M. Haertlein, M. Sferrazza, et al., Neutron reflectometry elucidates density profiles of deuterated proteins adsorbed onto surfaces displaying poly(ethylene glycol) brushes: Evidence for primary adsorption, Langmuir 29 (2013) 14178–14187.

CHAPTER FOUR

Role of the nanoparticle core and capping on the interaction with lipid monolayers

Martín Eduardo Villanueva[a,*], Santiago Daniel Salas[b], and Raquel Viviana Vico[b,*]

[a]Experimental Soft Matter and Thermal Physics (EST) group, Department of Physics, Université libre de Bruxelles, Boulevard du Triomphe CP223, Brussels, Belgium
[b]Instituto de Investigaciones en Fisicoquímica de Córdoba (INFIQC-UNC-CONICET), Departamento de Química Orgánica. Facultad de Ciencias Químicas, Universidad Nacional de Córdoba. Haya de la Torre y Medina Allende, Ciudad Universitaria, Córdoba, Argentina
*Corresponding authors. e-mail address: martin.villanueva@ulb.be; raquel.vico@unc.edu.ar

Contents

Abstract

The actual requirement of biotechnological tools to reduce the delivery time, cost, and animal assays in multiple pharmaceutical and (bio)medical applications has

Advances in Biomembranes and Lipid Self-Assembly, Volume 38
ISSN 2451-9634, https://doi.org/10.1016/bs.abl.2023.10.001

enhanced the historical search for rational design strategies to study the interaction of emerging nanomaterials with biological interfaces. Nanoparticles (NPs) are in increasing contact with different ecosystems. This raises the need to understand their interactions with biointerfaces, in order to prevent potential toxic effects and work on their rational design. In this sense, the modulation of the final response of NPs upon their interaction with biological fluids and interfaces requires deep knowledge of the properties of both their core and capping constituents and the extent to which they can determine the final fate of these interactions. This chapter presents Langmuir monolayers (LMs) as a potent tool for studying the interaction between different types of biologically relevant NPs and lipid model biomembranes. Herein we address the different experimental set up to study the films formed by NPs or NPs–lipids. Finally, we give insights into the potential and limitations of LMs as a high throughput technique to perform and validate in vivo studies, thus shortening the gap between experimental and clinical trials.

Abbreviations

A	molecular area.
A_0	limiting area.
A_C	collapse area.
AFM	atomic force microscopy.
AgNP-OA	oleic acid-coated silver nanoparticles.
AS	theoretical cross-sectional area.
AuNP	gold nanoparticles.
BAM	Brewster angle microscopy.
C_s^{-1}	surface compressional modulus.
DMPC	1,2-dimyristoyl-*sn*-glycero-3-phosphocholine.
DPPC	1,2-dipalmitoyl-*sn*-glycero-3-phosphocholine.
DPPG	1,2-dipalmitoyl-*sn*-glycero-3-phospho-(1′-*rac*-glycerol).
GIXD	Grazing incidence X-ray diffraction.
IOMNP	iron oxide magnetic nanoparticles.
IRRAS	infrared reflection absorption spectroscopy.
L	passivating layer length.
LC	liquid-condensed.
LE	liquid-expanded.
LM	Langmuir monolayer.
MNP-OA	oleic acid-coated magnetite nanoparticles.
MRI	magnetic resonance imaging.
OA	oleic acid.
NP	nanoparticle.
PC	phosphatidylcholine.
PDAC	poly(diallydimethylammonium chloride).
PEG	polyethylene glycol.
R	hard core radius.
Rhod-DOPE	*N*-(lyssamine Rhodamine B sulfonyl)-1,2-dioleoyl-*sn*-3-phosphatidylehanolamine.

SCM	stratum corneum mimic membrane.
SFG	sum frequency generation spectroscopy.
T	temperature.
TXRF	total reflection X-ray fluorescence.
X	molar fraction.
ΔG	free energy.
γ	interfacial tension.
Π	surface pressure.
Πc	collapse surface pressure.

1. Introduction

The increasing demand for nanoparticles (NPs) for biomedical, industrial, and technological applications has led to their spreading in the environment. NPs can now be considered emerging contaminants, and this emphasizes the requirement for knowledge about the extension in which the NPs composition determines their behavior toward biological molecules and natural barriers such as cell membranes [1,2]. As readily described by the concept *nano-bio interfacesace*, biological environments are constantly changing as a consequence of the different biomolecules and metabolites distributed in the system [3,4]. Intermolecular interactions and secreted substances easily affect medium conditions such as pH, osmotic pressure, ionic strength, and so on. Not to mention the so-called *protein corona* formation, in which proteins present in the biological fluid are associated and stabilized over the NPs surface as a result of a series of enthalpic and entropic interactions [5–7]. This will inherently change the surface identity of the NPs, thus affecting further interactions with biomembranes or receptor proteins residing in the plasma membrane of surrounding cells [8,9]. Therefore, when constructing NPs for biomedical applications, it is especially important to consider that this metastable state created by the biological medium will have a central role in the final NP–biomolecule interaction performance [3].

Given the above-mentioned conditions, several aspects must be considered regarding nanoparticle-controlled design. The accurate modulation of geometrical features will inherently impact the core functionalization and the interaction among neighboring particles or with biomembranes or other biomolecules [10]. Larger core surfaces will imply the availability of more sites for ligand functionalization, thus increasing the complexation rates with host biomolecules or receptors at the cell membrane. It is also worth considering that the many interfacial barriers comprising the different endocytic routes can be easily overcome for very low-sized NPs [10,11]. Therefore, not only the peripheral molecules decorating the NP structure but their inner *core* properties have consistently proved important in projecting many pharmaceutical and biotechnological applications, such as drug delivery and gene therapy. Nevertheless, the extent to which the *core* composition and physicochemical properties of functionalized NPs will affect the membrane properties remains poorly understood. In this sense, the constant search for accurate cell membrane models that can approach enough to reduce the gap between in vitro and in vivo studies still represents an issue for research on this topic.

Membrane models are simplified systems in which almost all physical and chemical parameters can be controlled. They are particularly useful for a systematic characterization of membrane attachment and disruption by NPs, and for visualizing NPs–membrane interactions. Apart from Langmuir monolayers, other membrane models are helpful for NPs-cell membrane interaction studies, such as lipid vesicles or supported lipid bilayers (SLB) and, more recently, computational modeling. Any of these models provide complementary data and allow the description of a broader picture of NPs–membrane interactions when used in combination. In the literature, there are outstanding reviews for a complementary reading regarding these other biomembrane models where the reader can find the advantages and limitations of each methodological approach [12–15]. In particular, Langmuir monolayers (LMs) have been extensively used with great success to study interactions in two-dimensional arrangements of amphiphilic molecules [16,17]. Being a simple physical model, they offer a precise control, at a fluid interface, over many experimental parameters, such as the available area per molecule at the air/water interface, the subphase composition e.g. simulating physiological conditions, and the control of temperature [18]. Thus, LMs represent one of the best choices in biomimetic models to understand the physicochemical interactions between NPs and biomolecules [19,20]. The strength of this approach resides in the possibility of systematically varying membrane composition to provide a hierarchical approach where interaction mechanisms can be determined in simpler models and then be extended to increasingly complex membranes.

The purpose of this chapter is to present a brief overview on the current knowledge of the effect that both the *core* and capping agents have on the interaction between functionalized NPs and Langmuir lipid films. To dissect the effect of the NP structural components, we established comparisons using nanoparticle systems bearing similar dimensions in which ideally only one variable is modified at a time (i.e., *core* or capping nature). Changes in membrane mechanical properties, lipid phase state and NPs preferential segregation, are some of the features that can been analyzed. We will start by describing the thermodynamics of Langmuir monolayers and complementary techniques to study interfacial phenomena, to then fully aboard our subject of study. Also, we will address the different experimental setups to study the films formed by NPs or NPs–lipids. In this regard, conclusions about the factors causing the NP–lipid interaction behavior have been given by firstly assessing the rheological properties of pure NPs films at the air/water interface. These examples are representative cases of study. Finally, we will highlight the scope of the LMs technique by considering the main key aspects regarding the biochemical events that NPs are prone to suffer in biological media, and how the mechanisms of these processes could be assessed with the current advancements in the LMs field. We will also contemplate the limitations of the LM approach to fully understand the impact of NP entity on the integrity of biomembranes, though remarking on the unique role of LMs as part of a multi-dimensional characterization of the phenomena taking place at the nanobiointerface.

The results presented here contribute to understanding the interplay between the structural components of NPs and lipid model membranes. Furthermore, comprehension of the NPs features that promote a determined change in cell membrane properties is of great interest for designing and developing nanoparticle platforms for many biotechnological applications.

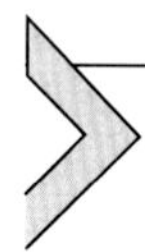

2. Langmuir monolayers as models of biological membranes

Langmuir monomolecular films have been extensively studied over many decades given the vast amount of detailed information that they provide from processes that occur at the membrane surface [17,20–23]. This simplistic method to form biomimetic systems allows precise control over the available area per molecule at the air/liquid interface, lipid composition, molecular packing, lateral pressure and several experimental parameters, such as composition or pH of the subphase, and temperature.

Basically, a defined number of amphiphilic molecules (e.g., a phospholipid or a mixture of lipids) from an organic solvent solution are deposited onto a fluid surface in a chemically inert trough (usually made by Teflon) by using a microsyringe. After a period, in general 10–20 min, in which the organic solvent is evaporated, the surface-active molecules will rearrange over the interface allowing their polar heads to remain in contact with the aqueous subphase while pointing their hydrophobic moieties towards the air. The trough is also equipped with movable barriers. When the barriers close, compression is exerted on the deposited molecular film while a microbalance measures the interfacial tension (γ) by the Wilhelmy plate method i.e., measuring the force on a hanging plate, usually made of chromatography paper or inert platinum, touching the surface. Given that the number of molecules deposited at the interface, as well as an accurate record of the mean molecular area, A (Å^2/molecule), are known with the corresponding surface pressure, Π (mN/m), it is possible to obtain a Langmuir isotherm [20–22]. In a Langmuir isotherm, the relationship of Π vs A is plotted. The shape of the isotherms can be correlated to the configuration of the molecules forming the monolayer. Thermodynamically, it can be explained by different two-dimensional analogs of different equations-of-state. Once the monolayer is compressed, the film can be allowed to expand by the opening of the movable barrier, permitting the study of hysteresis. This study allows to obtain relevant information on the thermodynamic and rheological behavior of the formed films at the interface [24]. Further details on these fundamentals and experimental procedures of these methodologies would be revisited in Section 3.

Recently, there has also been a developing interest in a different class of Langmuir monolayers composed, beyond the traditional amphiphiles, of inorganic NPs capped with different surfactants or polymers [25–27]. Similarly to amphiphilic molecules, these capped nanoparticles self-assemble into 2D lattices at the air/water interface. The stability of the NPs films greatly depends on the hydrophobic and hydrophilic properties of the NPs.

Additionally, the classical Langmuir trough setup can be coupled with a wide range of techniques to broaden the information on the interfacial phenomena. Brewster Angle Microscopy (BAM) is a powerful tool with a resolution in the mesoscale range that allows for real-time visualization of Langmuir films. It is a label-free optical technique which also does not require any additional sample modification allowing precise imaging while maintaining the biophysical and physicochemical properties of the samples [28]. The Langmuir balance can also be coupled to several other techniques, such as Infrared Reflection Adsorption Spectroscopy (IRRAS), Sum Frequency Generation Spectroscopy (SFG), Grazing Incidence X-ray Diffraction (GIXD), and Total Reflection X-ray Fluorescence (TXRF), among others [20]. The combination of these different methods allows for an extensive mechanistic study of the interactions between NPs and lipid model membranes.

3. Thermodynamics of Langmuir monolayers

The interfacial response of a model membrane composed by lipid amphiphiles during monolayer compression is a direct consequence of the free energy excess experienced by water molecules at the surface compared with those in the bulk subphase [29,30]. For two-dimensional systems at constant number of molecules, this is given by the differential of the Helmholtz free energy equation:

$$dF = -SdT - \Pi dA \tag{1}$$

Where F correspond to the Helmholtz free energy, S represents the surface entropy along a temperature differential dT, and Π the surface pressure along an area differential dA. So, at constant temperature:

$$\left(\frac{dF}{dA}\right)_T = \gamma \tag{2}$$

Where γ is defined as the interfacial tension. From a mechanical point of view, γ is a force parallel to the interface per unit length, thus a two-dimensional analog of pressure. For the bare air/water interface the surface tension is γ_0. In equilibrium, the presence of a floating monolayer reduces the interfacial tension that is characterized by γ. The resulting Π is expressed as Eq. (3) in analogy to the bulk osmotic pressure [29,30].

$$\Pi = \gamma_0 - \gamma \tag{3}$$

In Langmuir monolayers, the surface pressure Π can be plotted as a function of the average molecular area (A) of the film at constant temperature, while movable barriers reduce the available trough area. As previously introduced, the established Π vs A relation will give an isotherm whose behavior resembles the ideal gas equation that relates the pressure P and the volume V of a simple fluid. Therefore, it is possible to apply similar assumptions to explain the mixing behavior of the different surface-active molecules interacting at the air/aqueous interface. For an ideal binary mixture of two amphiphiles the mean molecular area A_{12}^{id} at given surface pressure, can be calculated from the pure components isotherms as,

$$A_{12}^{id} = [A_1 x_1 + A_2 x_2]_{\Pi} \tag{4}$$

Where A_1 and A_2 denote the mean molecular areas at a given surface pressure in the one-component films and x_1 and x_2 are the mole fractions of components 1 and 2 in the mixed films, respectively [31].

The different regions of an isotherm are generally distinguished through terminology associated with each phase state. Briefly, the molecules are in a diluted regime similar to the gas phase at large areas. As the area decreases because of the barriers closing, the system lit off from the gaseous phase and becomes condensed (liquid or solid), and Π increases rapidly. As a result, it is possible to elucidate the interaction mechanisms taking place between the amphiphilic molecules in terms of their physicochemical properties and the forces taking place during their interaction [29,30,32]. As we shall see in the following sections, the many lipid phases that can be distinguished from the discontinuities observed in the isotherms will also give valuable information on the main forces developed at the biointerface (i.e., van der Waals or electrostatically driven interactions), given the high sensibility of the LMs technique to sense these phenomena. From a thermodynamic point of view, the type of possible interactions can be determined from the excess area per molecule (A_{exc}), at a given surface pressure, as for the example given by Eq. (5), it can be written as:

$$A_{exc} = [A_{12} + A_{12}^{id}]_{\Pi} \tag{5}$$

Where excess areas equal to zero will indicate either ideally miscible or totally immiscible films. At the same time, negative or positive deviations will hint at possible attractive and repulsive intermolecular interactions, respectively. The magnitude of these interactions can be evaluated through the excess Gibbs energy of mixing, ΔG_{exc}, where N is Avogadro's number [30,31].

$$\Delta G_{exc} = N \int_0^{\Pi} A_{exc} \, d\Pi \tag{6}$$

Subsequently, the thermodynamic stability of the mixed systems can be characterized based on the total Gibbs energy of mixing, ΔG_{mix}:

$$\Delta G_{mix} = \Delta G_{exc} + \Delta G_{id} \tag{7}$$

Where the ideal Gibbs energy of mixing, ΔG_{id}, can be expressed as: [30,31].

$$\Delta G_{id} = RT\,(x_1 \ln x_1 + x_2 \ln x_2) \tag{8}$$

With R being the gas constant and T the temperature.

Another way to characterize the surface-active molecules' intermolecular interactions relies on performing successive compression–decompression cycles of the formed Langmuir films. Just before the collapse surface pressure (Π_c), the molecules still form a monomolecular film, ideally aligned in close contact. Thus, reversible and irreversible interactions forced by the film compression can be equally considered prone to happen at this stage. Depending on the nature of these interactions (e.g., chain interdigitation, complexation, electrostatic interactions, etc.), the system could return over the same path to the initial energetic state after barrier decompression. In other words, the isotherm described by the monolayer expansion allows us to determine the amount of energy in the form of compressional work the system can store or release after its molecules are subjected to the minimum close contact interaction [29]. This is related to the hysteresis behavior of the system, and its thermodynamic functions can also be associated with the ideal mixing regime as described before [30]. Ideal mixing shows no hysteresis, so $\Delta G_i^{hys} = 0$, $\Delta S_i^{hys} = 0$, and $\Delta H_i^{hys} = 0$. Then, for the experimental compression–expansion isotherms of a mixture, the free energy of hysteresis ΔG_m^{hys}, the configurational entropy of hysteresis ΔS_m^{hys} (from which the entropic contribution to the free energy of hysteresis $T\Delta S_m^{hys}$ can be obtained), and the enthalpy of hysteresis ΔH_m^{hys} are defined by Eqs. (9)–(11), respectively [29,30]:

$$\Delta G_m^{hys} = \Delta G_{expansion} + \Delta G_{compression} \tag{9}$$

$$\left[\Delta S_{\Pi}^{hys} = R \ln \frac{A_{expansion}}{A_{compression}}\right]_{\Pi,X} \tag{10}$$

$$\Delta H^{hys} = \Delta G^{hys} + T\Delta S^{hys} \tag{11}$$

On the assessment of rheological and mechanical properties of the formed amphiphiles or lipids films, the surface compressibility of the film (C_S) is analyzed. In practice, the reciprocal expression of C_S is more convenient, the surface compressional modulus (C_S^{-1}), where C_S^{-1} has dimensions of N/m. The surface compressional modulus C_S^{-1} provides a quantitative measure of lipid in-plane packing elasticity, and is used to facilitate comparison with elastic moduli of area compressibility values in bilayer systems [30].

$$C_S^{-1} = -A\left(\frac{d\Pi}{dA}\right)_T \tag{12}$$

It can be seen from Eq. (12) that for films in which ΠA = constant, C_S^{-1} will tend to Π. Thus, for a clean surface, the modulus C_S^{-1} is zero, increasing with the amount of surface-active material present. In general C_S^{-1} will depend also on the state of the film, being greater for more condensed films [30]. The characterization of obtained experimental compressibility can also be studied by establishing comparisons with the ideal compressibility modulus ($\overline{C_S^{-1}}$), taking in count the individual C_S^{-1} for each component in the mixture, occurring at a certain fraction of molecular area (A_n) at a given mole fraction (X_n) [33–35]. Thus, at a given surface pressure:

$$C_S^{-1} = \left(X_1\left(\frac{C_{S\ 1}^{-1}}{A_1}\right)_\Pi + X_2\left(\frac{C_{S\ 2}^{-1}}{A_2}\right)_\Pi\right)(X_1A_1 + X_2A_2)_\Pi \tag{13}$$

C_S^{-1} provides vital information on the lateral packing elasticity (easiness or resistance to lateral compression) inside the monolayer. Moreover, C_S^{-1} values are highly sensitive to the sudden alterations in the lipid structures during the lateral interaction with hard lipids such as cholesterol [30].

Regarding the importance of having reliable biomembrane models to validate biomedical applications, although bidimensional configuration of LMs cannot geometrically resemble the bilayer configuration of lipid vesicles, a direct monolayer–bilayer correspondence can be established in order to *homologate* the thermodynamic conditions of both models. Eq. (14) defines the requirement for the monolayer surface pressure (Π_m) at a certain area per lipid molecule occupied in a bilayer arrangement (a_b^0) to achieve the correspondence conditions [36].

$$\Pi_m(a_b^0) = \gamma_{phob} + \left[\phi'_{m-m}(a_b^0) - \phi'_{m-u}(a_b^0)\right] \tag{14}$$

Where γ_{phob} is defined as the hydrophobic surface free energy density; and ϕ'_{m-m} and ϕ'_{m-u}, the free energies of interaction between lipid chains of opposing monolayers in bilayer arrangements, and between the lipid chains and the upper phase (i.e., air or oil) at monolayers, respectively. A fundamental requirement for correspondence between the two models is that the area per lipid molecule in the monolayer should be the same as that in the bilayer, i.e., a_b^0. If the second term on the right in Eq. (14) is zero, the lipids in monolayers and bilayers may be in closely similar states, since both are then subject to the same compressive tension, which for bilayers has a value that would be numerically comparable to γ_{phob}. Given the experimental conditions in which LMs can be set up, this condition is likely to be met for monolayer at the oil–water interface but, for monolayers at the air–water interface, only if the system satisfies the condition of $\phi'_{m-m} \ll \gamma_{phob}$. In this sense, a closer approach for validating correspondences in LMs at the air/water interface can be given for monolayers at an average surface pressure of ~30 mN/m (Π_m). [36] At this Π_m it is postulated that the packing density of lipid molecules is found to be similar to the internal pressure of cell membranes thus providing more accuracy on the equivalences established. In sum, these conditions represent an important finding since they mean that monolayers allow to explore wider regions than those limited by the lipid assembly state equation, from the obtained Π vs A isotherm, in comparison with experiments with tensioned bilayers [36].

Additionally, it is possible to take advantage of the LMs set up to perform studies on amphiphile adsorption or penetration into a bare interface or on a preformed lipid monolayer packed to a specific initial surface pressure (Π_0) [35,37,38]. In this experiment, the trough area is kept constant and the amphiphiles (or NPs) are injected into the subphase; in the case of amphiphiles, the concentration into the subphase has to be well above the critical aggregation concentration, and for NPs, these need to be dispersible in the subphase. Therefore, for NPs, the dispersibility and kinetic colloidal stability in water or the solution used as the subphase need to be evaluated before adsorption or penetration studies. The Π is registered as a function of time until a constant surface pressure is reached (Π_f); this lets to know the change in surface pressure as a result of the continuous incorporation of molecules (or NPs) from the bulk of the aqueous subphase into the bare interface or the preformed lipid layer. These experiments are referred as Gibbs adsorption isotherms and the change in surface pressure is calculated according to Eq. (15):

$$\Delta\Pi = \Pi_f - \Pi_0 \tag{15}$$

Where Π_0 is the surface pressure of the bare interface or preformed lipid monolayer and Π_f the final surface pressure after reaching molecular (or NPs) saturation at the air/water interface [i.e., no additional molecules (or NPs) are incorporated into the bare interface or lipid monolayer]. Therefore, the obtained $\Delta\Pi$ can provide information on the ability of a certain diffusing amphiphile or NP to incorporate into the biomembrane having a particular packing organization. In most examples in the literature, the plot $\Delta\Pi$ vs Π_0 is linear. Thus, the critical surface pressure (Π_{cr}), where $\Delta\Pi$ is zero and no molecules (or NPs) are incorporated into the lipid monolayer, can be extrapolated from the $\Delta\Pi$ vs Π_0 plot [35,37,39]. Another possibility offered by the Langmuir balance is to evaluate the changes in molecular area at a selected surface pressure value; in this manner it is possible to evaluate the film stability and the squeezing of the film constituents (i.e., if molecules or NPs are desorbed from the interface to the subphase). These approaches find interesting applications for instance on the validation of in vitro studies for drug delivery and cell penetration. We will discuss the different experimental approaches in Section 5.

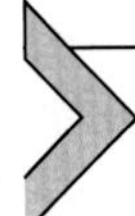

4. Self-assembly of nanoparticles at the air/water interface

The behavior of NPs at liquid interfaces is not only of fundamental interest concerning the understanding of NPs interaction with biological membranes. Also, this knowledge is essential for understanding the physics of two-dimensional systems formed by NPs. It is important for practical applications, such as stabilizing emulsions and foams, colloidal self-assembly, and optoelectronic devices [38–40]. Several experimental studies of NPs at interfaces have reported the formation of two-dimensional structures motivating new theoretical approaches to colloidal interactions beyond the traditional bulk considerations.

The fate of the NPs at the air/water interface depends on the chemical nature of the particle and the chemical nature of the phases adjacent to the interface [40]. These features will define the contact angle (θ) between the nanoparticles and the interfacial plane and, thus, the free energy of adsorption at such interface. The contact angle of particles is analogous to the hydrophilic–lipophilic balance (HLB) of typical surfactants; thus, the partition of the particles between the two fluids phase is correlated to θ. The whole picture considers the perimeter of the three-phase contact line (gas/liquid/solid), which also depends on the developed line tension (τ) among the phases. Although the effect of line tension is negligible in the case of microparticles, it can be significant for nanoparticles in which the line tension may lead to their detachment from fluid interfaces. Therefore, the free energy of adsorption of particles at the nanometer scale will depend on these variables according to the following equation defined elsewhere [40]:

$$\Delta E_p = \gamma_{f1f2} \cos\theta_{\infty} 2\pi R^2 (1 - \cos\theta) + 2\tau\pi R \sin\theta - \pi R^2 \gamma_{f1f2} \cos^2\theta. \tag{16}$$

Where γ_{f1f2} is interfacial tension between the fluid phases, and θ is related to the contact angles (being θ_{∞}, the macroscopic contact angle as defined by Young's equation) [41]. As can be seen, the free energy of adsorption also considers the particle radius (R). Dependencies with nanoparticle size have also been widely explored. In this sense, larger nanoparticle sizes result in higher contact angles given by the large decrease in surface tension. The broad number of techniques that have been used to estimate the θ of NPs of different sizes, has largely contributed to increase the margin of error among the many reports that can be find in the literature. Therefore, to define a certain size threshold for the NP adsorption/desorption or their most probable position at the air/water interface, remains rather than inaccurate. Despite all that, the most important criteria for particle adsorption would be given by the important contributions of the different intermolecular forces to the surface tension, including van der Waals, polar, and electrostatic interactions. Neither of these theories considers specific densities or weight-related properties of the particles. Once again, at the nanometer scale, intermolecular forces remain way more important to define their stability at the air/water interface. For instance, London–van der Waals (dispersion) interactions rely on terms of the Hamaker constants of the three materials. The latter concomitantly depends on the density number of particles occupying a certain area and is also related to the dielectric permittivity (ε) of the three materials (phases), among other parameters. This brings us to one more parameter that can determine the adsorption equilibrium of the NPs, and it is related to the surface charge. If one of the different media is conducting, e.g., metal or semiconductor nanoparticles, it can lead to repulsion/agglomeration phenomena depending on the charge sign the NPs are exposing. As we have pointed out previously, capping ligands can be a very helpful option when it comes to NP stabilization at the air/water interface. Still, even in these cases, the chemistry of capping ligands should be

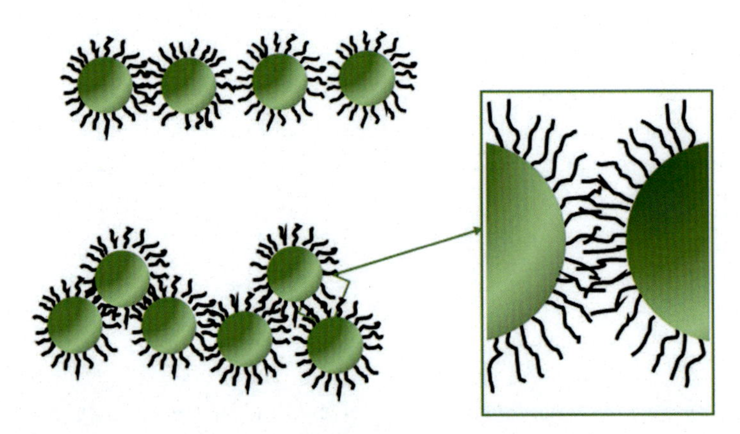

Fig. 1 Schematic representation of interdigitated nanoparticles.

thoroughly studied to overcome side issues, such as charge interactions, dissociation, etc. [40,42].

Nowadays, NPs are considered the new building block of advanced materials that are expected to exhibit unusual chemical and physical properties. On the facing of the challenges of designing new NP platforms, it be must anticipated that, the collective properties of such mesoscopic assemblies will differ from isolated NPs and the bulk phase [43,44]. As mentioned above, NPs self-assembly at the air/water interface is a process governed by an interplay between many interparticle interactions, but also, by interactions among the capping molecules present at the NPs surface. The later, being responsible of generating an interpenetration phenomenon generally called, interdigitation (Fig. 1). Different studies have demonstrated that the establishment of interdigitation, in many cases, dictates the behavior of the NPs film [31,45] or 2D, 3D NPs arrays [46].

4.1 Sterically-stabilized neutral nanoparticles

Sterically-stabilized NPs occur with hydrophobic NPs, or partially hydrophobic NPs capped with molecules with an alkane moiety [47]. Such passivated NPs form arrays at the air/water interface where the ratio of the hard (i.e. metal) *core* radius (R) to the passivating layer length (L), to a great extent, determines the morphology of the self-assembled structures [48]. In this regard, Gelbart et al. [49] proposed that increasing the nanoparticle concentration drives the formation of stripes from circles, and networks from linear chains depending on the relationship between R and L. For a small ratio of passivating layer length to the core radius ($L < R$), circular islands are favored, whereas for a large ratio ($L > R$), stripes are formed. In the latter case, the anisotropy needed to create chains is commonly assigned to interdigitation effects. In line with experimental results, simulation studies confirm the dependence of the self-assembled structures on the ratio of passivating layer length to *core* radius. At low packing fractions, nanoparticles passivated with short surfactant layers (butanethiols) form open structures with a local ordering corresponding to hexagonal packing. For the same core size, longer surfactants (dodecanethiol) give rise to more compact structures exhibiting a change in the crystal symmetry from hexagonal to a distorted square lattice [47,49].

4.2 Charged nanoparticles

When working with Langmuir monolayers of charged NPs, it is found that charged NPs, unlike hydrophobic or amphiphilic ones (i.e. capped with block copolymers), form less stable two-dimensional structures at the air/water interface because they are more susceptible to desorption from the interface. Volpe Bossa et al. found that charged particles at the air/water interface are subject to long-range dipolar interactions [50]. However, the description of the physical properties of the arrays formed by these NPs is challenged by the complexity of electrostatic interactions near and across discontinuous dielectric interfaces. The interaction between two charged particles trapped at the air/water interface becomes dipolar at large separations. The corresponding dipole moment can be modeled by considering a single-point charge located exactly at the interface. However, this model still fails to correctly predict the dipole moment's dependence on the salt concentration in the aqueous medium [50]. Pioneering work performed by Pieranski [51], who studied charged polystyrene microparticle assembly at the water liquid–vapor interface, reported electrostatic repulsions between particles whose range is longer than the electrostatic repulsions in bulk water. This is due to the electric dipoles formed by the colloid surface charge and the counterions, resulting in a power law distribution of the dipole–dipole repulsion between the particles, which drives the formation of compact 2D lattices at air/water interfaces. Subsequent work from Aveyard et al. [52] also using polystyrene particles showed the sensitivity of the monolayer structures to electrolyte concentration (water salinity) [52]. At low salinity, the repulsion between particles results in fairly ordered arrays. Increasing the salt concentration appears to screen the repulsive interactions and favors irreversible 2D cluster formation, similar to what accounts for colloidal dispersions of charged NPs.

4.3 Magnetic nanoparticles

Magnetic NPs attract the attention of the technological and scientific community and are the subject of intensive studies [53]. As with charged NPs, the interactions present among magnetic NPs are rather complex. In magnetic particles, the presence of magnetic dipolar interactions can be considered. Iron oxide nanoparticles with a radius of about 5 nm will be superparamagnetic. So the magnetic dipolar force between them should be zero, but this will not be the case if an external field magnetizes them [53].

Lefebure et al. [54] showed the compression of hydrophobic magnetic NPs films results in the formation of circular domains as well as close-packed arrays for high surface pressures. Experiments with similar magnetic nanoparticles, γ-Fe_2O_3 (7.5–15.5 nm in diameter) at the air/water interface have shown the formation of chains and compact circular aggregates reminiscent of the aggregates observed in neutral and charged particles. These experimental observations have been interpreted in terms of a balance between van der Waals and magnetic dipole–dipole interactions [54].

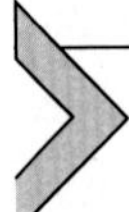

5. Experimental approaches to study nanoparticles and nanoparticles–lipids at the air/water interface

The possible experimental setups that can be adopted to evaluate the effects of NPs on lipid monolayers can be categorized mainly into three distinctive approaches, as shown in Fig. 2A depicts the setup for obtaining the classical Langmuir isotherm. First, the subphase (i.e., water, buffer solution) is placed at a constant temperature in a Langmuir trough equipped with movable barriers. Then, an organic lipid solution or a dispersion of any hydrophobic NP is spread at the interface. Here it is important to mention that the concentration of the spread lipid solution or NP dispersion must be well known to obtain the Langmuir isotherm in terms of an area per molecule or per particle. Also, this setup allows the study of different lipids mixtures or lipid–NPs mixtures. In this case, preparing a mixed solution of different lipids or lipid/s and NPs and spreading the mixture at the interface is more practical. This procedure is the well-known co-spreading methodology to obtain mixed monolayers. In the case of hydrophilic or amphiphilic NPs, their combination with lipids by using adequate solvents (in general, chloroform mixed with methanol or other polar solvents) also permits the co-spreading of the mixtures at the air/water interface. The interfacial stability of the spread lipid, NP, or their

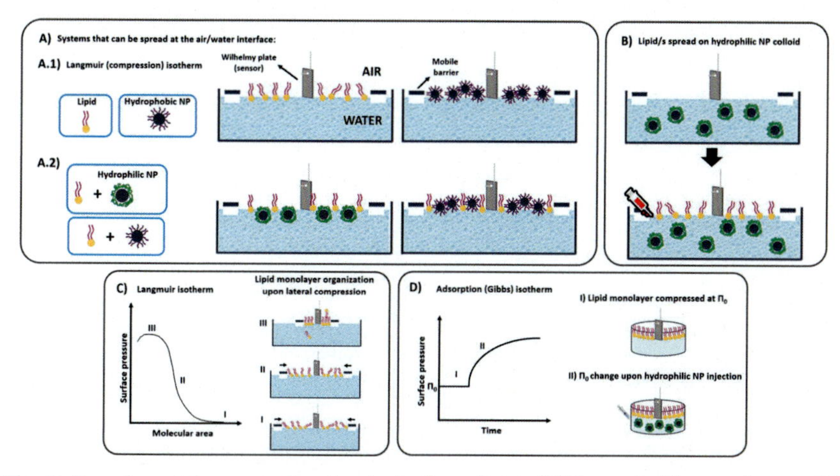

Fig. 2 Experimental approaches to study the effect of NPs on lipid monolayers. (A) Spread and formation of Langmuir films made of lipids or hydrophobic NPs (A.1) or mixtures of hydrophobic or hydrophilic NPs and lipids at the air/water interface (A.2). (B) Spread and formation of lipid monolayers over a colloidal dispersion of hydrophilic NPs. (C) Different stages of lateral compression of a lipid (NP or mixed) film (I, II, III) and acquisition of the Langmuir isotherm. (D) Adsorption (Gibbs) isotherm obtained by injection of hydrophilic NPs in the subphase of a preformed lipid monolayer.

mixtures can also be evaluated. One option is to assess the change in area over a certain period of time when a monolayer is compressed and maintained at a particular surface pressure; this is done by automatically closing or opening the mobile barriers (change of area). In the case of loss of monolayer components, a closing of the barriers will be observed to keep constant the surface pressure. If the monolayer is stable, its area must remain constant in time. If desorption of material at the air/water interface occurs, the decreasing surface pressure will be compensated as a reduction of the surface area [55,56]. Another option to test the loss of material from the interface is performing a compression–decompression cycle and a second compression. Upon the second compression, the recovery of the surface area at a particular surface pressure near the collapse indicates that no material is lost from the interface. If the surface area is smaller after the second compression, this could mean that desorption of the material is taking place [31]. Desorption of material from the interface into the aqueous subphase can also be verified by carefully removing the monolayer from the interface (e.g. by transferring the monolayer to an adjacent trough with the help of a barrier) and recovering the subphase [35]. The subphase can then be lyophilized, if required, and analyzed through any analytical technique. These approaches allow us to draw conclusions based on the average interfacial behavior by comparing the compression isotherms of the mixtures with those of the lipid/s and NPs monolayers and their stabilities. However, discerning specific interactions of lipid molecules with single or clusters (agglomerates) of NPs is limited. This is why, in most cases, the Langmuir technique is used concomitantly with other complementary surface characterization tools, such as AFM, BAM, IRRAS, SFG, among others [14,15].

The interfacial behavior of hydrophilic NPs also can be studied by preparing a dispersion of a known concentration of NPs and using it as the subphase. Once the NPs dispersion is placed in the Langmuir trough, the lipid/s solution is spread at the interface, Fig. 2B. After evaporation the organic solvent, the compression can be started. In this way, and depending on the surface properties of hydrophilic NPs, these can be anchored at the air/water interface by interacting with the lipid/s molecules due to favorable interactions established among them. Once the compression is performed, the isotherm obtained can be compared with the isotherm of the lipid/s obtained from a subphase without NPs. In this case, as it is impossible to know the amount of adsorbed or interacting NPs at the interface, the area can be expressed as a function of the area of lipid/s molecules. The arrival of NPs at the lipid interface resembles the biological event in which the NPs interact, penetrate, or embed into the cell membrane.

The third approach, Fig. 2C, also applies to the study of hydrophilic NPs and allows to obtain adsorption isotherms of the NPs to a bare or a preformed lipid monolayer present at the interface and packed to a particular surface pressure. Here, the NPs are added to the subphase through a lateral injection port for not disrupting the lipid monolayer that was previously seeded at the interface. Although it is necessary to stir the NPs into the subphase to facilitate homogenization, this cannot be performed with magnetic particles because of particle attachment to the magnetic stirrer. Thus, sufficient time prior to measurement must be allowed for homogeneous particles' diffusion. In this experimental approach, the ability of the NPs to reach and interact with a preformed lipid monolayer present at the interface is evaluated. For this, the NPs need to overcome interactions with other particles and the aqueous solvent and reach the air/water interface. This scenario resembles what would occur, for example, with NPs in the extracellular matrix approaching a cell membrane.

Regarding the adsorption of hydrophilic NPs to a bare interface, it has been observed that many hydrophilic NPs cannot form stable Gibbs films on their own. However, if a lipid monolayer is present at the interface, the hydrophilic NPs exert measurable changes in the surface pressure on time [57–59].

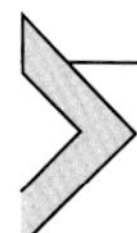

6. Case study: Hydrophobic, amphiphilic, and hydrophilic nanoparticles at the air/water interface

NPs stability at interfaces depends upon a balance between long and short-range interparticle attractive and repulsive forces [49,60]. It is also known that NPs films, upon compression, undergo transitions from monolayers to multilayers and can buckle, wrinkle, or fold out of plane similarly to traditional surfactants [44,49]. In a big picture, the hydrophobicity would likely define the NP stability at the air/water interface. Basically, if the NPs are not hydrophobic enough or amphiphilic (i.e. capped with a block copolymer), they will sink into the subphase, whereas too much hydrophobic coating can cause aggregation and stacking [61,62]. However, the hydrophilic/hydrophobic balance is not the only factor affecting the 2D patterns NPs acquire at fluid interfaces. For magnetic NPs has also been seen the *core* plays an important role, especially in the mesoscale range [31]. Currently, there are few studies about the effect of the inorganic *core* composition of NPs in terms of the properties of the 2D arrays that NPs can form in fluid interfaces. Nevertheless, all these factors influence the 2D ordering of NPs and dictate the mechanical response and subsequent collapse behavior of assembled monolayers.

Concerning the capping and *core* effect on NPs assembly, here we address some case studies of hydrophobic, amphiphilic and hydrophilic NPs at the air/water interface and the surface properties and 2D arrays they form. We will assess the extent to which hydrophobic NPs bearing different core compositions, but functionalized with the same surfactant can present significantly different behaviors. For this, we present of oleic acid (OA) capped magnetite (Fe_3O_4) and silver NPs. Also, we will revise some relevant cases from the bibliography in which the length of the hydrocarbon chain of alkanes has been modified, but the same *core* has been preserved. Regarding NPs with hydrophilic character, their films are less stable, as mentioned above, but some common characteristics are presented from work in the literature. The study of the interfacial properties of

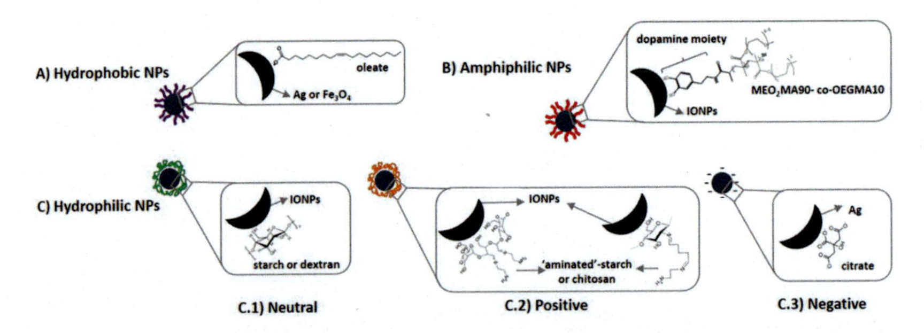

Fig. 3 Schematic representation of some functionalized nanoparticles along with the structure of their capping agents.

hydrophilic NPs, even if they do not form films, is interesting to evaluate the synergic effect between the *core* and the capping that modify the properties of lipid monolayers. Fig. 3 shows a schematic representation of some of the NPs and their capping molecules that will be mentioned in the preceding sections to facilitate interpretation.

6.1 Hydrophobic nanoparticles

Here we revisit some of the most relevant findings on the interaction among hydrophobic NPs. To begin with, we present our data on oleic acid-coated silver and magnetite NPs, namely AgNP-OA and MNP-OA, respectively. These nanoparticles form highly stable films at the air/water interface. On one hand, AgNP-OA bearing an Ag^0 silver core was synthesized by thermal reduction following the protocol cited in [45]. A size of about (5.2 ± 0.1) nm and a strong plasmon resonance absorption band around 412–414 nm were confirmed for these NPs. On the other hand, MNP-OA bearing a Fe_3O_4 magnetic core was obtained by co-precipitation as described elsewhere [31,45]. The average diameter was determined as (7.1 ± 0.7) nm, and superparamagnetic properties were obtained for these NPs. Hence, both NP systems were seen to develop structural and physicochemical properties that are highly sought by many biotechnological applications. On top of this, the oleic acid molecules attached to the NPs cores by chemisorption (mainly bidentate chelation through the carboxylate), formed a single layer that confers to the particles hydrophobic character and high interfacial stability thus, making them perfect candidates for Langmuir monolayer studies.

Fig. 4 depicts the isothermal behavior of MNP-OA and Ag-OA at the air/water interface. As shown, both hydrophobic nanoparticles could form

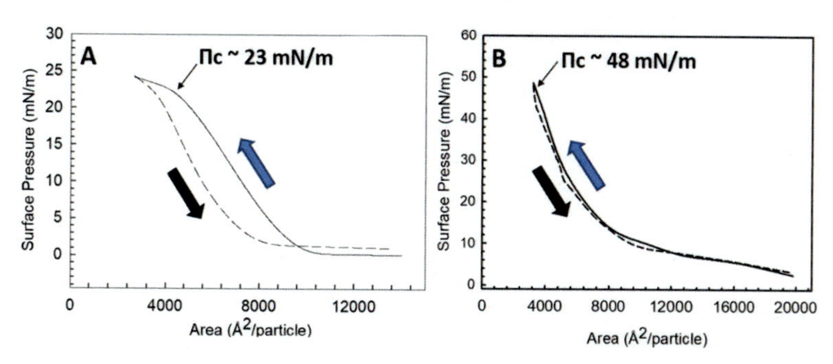

Fig. 4 Variation of the surface pressure with the average area at the air/water interface for pure MNP-OA (A) and AgNP-OA (B) Langmuir monolayers. Compression (solid line) and expansion (dashed line). The Collapse pressure (Π_C) of each monolayer is indicated in the plots.

highly stable films under successive compression–decompression cycles. Nevertheless, the most interesting differences arise from the hysteresis thermodynamics [63,64]. An increasing degree of hysteresis was detected from the compression–decompression cycles imposed to the MNP-OA (Fig. 4A). As calculations reflected, the large negative value obtained for ΔG^{hys}, indicated the capacity of the magnetite nanoparticle films to store energy in the form of organizational information [31]. Regarding AgNP-OA, the observed degree of hysteresis was nearly zero indicating that the system does not retain energy that could favor, for example, new particle–particle interactions (Fig. 4B). The energetics involved in those differential organizational arrangements taking place at the inter-particle interactions is not a trivial matter and it would be logical to consider their intervention on the observed response when nanoparticles interact with phospholipids. To elucidate this phenomenon more quantitatively and understandably, we established comparisons using parameterized areas obtained from the isotherm plots (Table 1).

As depicted in Table 1, the collapse areas (A_C) obtained for both films of pure NPs were lower with respect to their respective limiting areas (A_0). Along with the optical density of the film which topographies were studied by BAM, the difference between the calculated areas, suggested that a multiplicity of inter-particle arrangements could be taking place when the nanoparticles reach the highest compaction states. The relation between the theoretical cross-sectional areas (A_S) and the A_0 areas, was found to be close to 1:1 and 1.5:1 for AgNP-OA and MNP-OA, respectively. These observations regarding the interfacial behavior of both NPs, were further

Table 1 Parameterized areas in Å^2/particle. A_C was obtained from the NPs isotherms at the collapse pressure point. A_0 areas were obtained by extrapolating the linear part of the isotherms at zero surface pressure. Theoretical cross-sectional areas were determined from the structural characterization of the NPs.

Hydrophobic NPs	Collapse area (A_C)*	Limiting area (A_0)	Theoretical cross-sectional areas (A_S)
MNP-OA	5147	9241	13,260
AgNP-OA	2212	6300	5945

*Error in area at Π_C is lower than 0.1% for both nanoparticles.

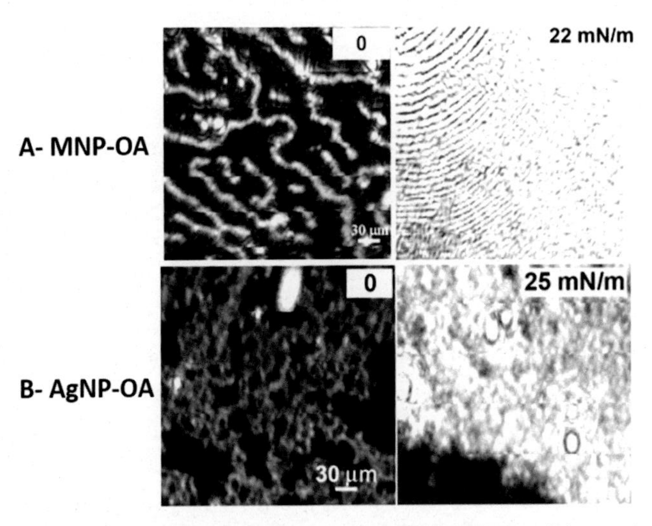

Fig. 5 BAM micrographs of pure MNP-OA and AgNP-OA at the air/water interface at the onset of lateral compression (~0 mN/m) and after reaching high packing states at ~20 mN/m.

complemented with BAM microscopy. The BAM images shed light on the organization at the mesoscale level as well as the textures of the surfaces (Fig. 5).

On one hand, oleic acid ligand interdigitation [65] between nanoparticles explained the observed results in MNP-OA. Consequently, magnetic nanoparticles formed long-range ordered structures (wire-like) that were seen to be preserved along the entire isotherm. On the other hand, AgNP-OA were found to maintain an in-plane interaction until averaged surface pressures of 12–15 mN/m (A_S ~ 6000), without establishing interdigitation processes. Beyond that point, the AgNP-OA organized in a 2D array [66] with an aleatory segregation pattern up to near the collapse of the film. The degree of hysteresis observed is then clearly related to the amount of work that is needed for each nanoparticle to reorganize the film structure upon film decompression. Both nanoparticles differed from each other in the amounts of energy required to return to the initial state (depending on the inter-particle interaction strength), and resulted to show interesting behaviors once interacting with lipid films as will be shown in the following section.

An interesting comparison about the properties of Au and Ag NPs capped with dodecanethiol in Langmuir monolayers was presented by You et al. [67]. In this work, they compared the Langmuir films properties of commercially available Au-NPs (5.1 ± 1.2 nm) and Ag-NPs (5.5 ± 2.5 nm) ligated with dodecanethiol with other characterization techniques like AFM and optical microscopy. The isotherms and optical images indicate that the Au and Ag nanoparticles undergo a similar series of morphological transitions upon compression, differing mainly in their mode of collapse. For both NPs were observed at the beginning of the barrier compression that each nanoparticle sample consists of numerous particle rafts separated by void space of open air/water interface, similar to that described above for MNP-OA and Ag-OA. Upon further compression, these rafts begin to merge, causing a steep increase in surface pressure and a corresponding increase in shear and compressive moduli. At this point, the images obtained by optical microscopy showed optically uniform monolayers [67]. The hysteretic behavior of these dodecanethiol-capped NPs was not addressed to make a comparison between NPs with a different capping than OA and with an Au *core*. Carrying out these kinds of comparisons would be of great value to know whether only magnetic NPs are capable of storing organizational information once they are compressed [31].

6.2 Amphiphilic nanoparticles

Amphiphilic synthetic polymers, such as polyethylene glycol copolymers and poloxamers have been employed as stabilizing agents of iron oxide MNPs (IOMNPs) [63]. In this case, the particles were endowed with the characteristic hydrophilicity and surface-activity of their coatings and were able to form Langmuir films as stable as the polymers alone [68]. Similarly, hydrolyzed poly(maleic anhydride-*alt*-1-octadecane) as the coating agent of AgNPs rendered amphiphilic carboxylate-decorated particles with surface activity due to hydrophobic interactions at the bare air/water interface. Furthermore, the amphiphilic carboxylated-AgNPs were coated on top with polyethylenimine (PEI) in order to revert the surface charge to positive and the particles still presented surface activity owing to the presence of the alkyl spacer [69]. Similar information to that obtained for hydrophobic NPs can be accessed for amphiphilic NPs.

6.3 Hydrophilic nanoparticles

Contrary to hydrophobic NPs, the greater aqueous dispersibility of hydrophilic NPs renders them unable to form stable films at the air/water interface as they tend to sink rapidly in the aqueous subphase. This behavior is mostly governed by the polar nature of the capping and is independent of the core identity. Coated NPs, such as IOMNPs functionalized with citrate [70], neutral polysaccharides, [58,71], amine-derivatized polysaccharides, [59,72], as well as AgNPs coated with citrate [57] or polyether block polymers amide [73] all shared the same incapability to adsorb and remain at the bare air/water interface.

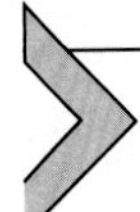

7. Case study: Interaction of hydrophobic, amphiphilic, and hydrophilic nanoparticles with lipid model membranes at the air/water interface

Lipid layers are considered among the first protective barriers of the human body (i.e. skin, lung surfactant, or tear film) against emergent NPs pollutants; also, they represent the target for or an obstacle for many NPs designed for theranostics. This makes exploring the physicochemical bases underlying the interaction of NPs and lipid layers necessary [74]. Furthermore, particle incorporation worsens the mechanical performance of lipid layers, which may negatively impact different processes presenting biological relevance. The modification induced by the NPs is dependent on their specific chemical nature. In this section, we shed light on some of the most fundamental physicochemical bases governing the interaction of NPs with lipid layers, which plays an essential role in designing strategies for optimal NPs activity when required or preventing the potential health hazards associated with NPs pollution.

In the following section, we will recount some experimental findings reported on the interaction of lipid Langmuir monolayers with NPs coated with hydrophobic, amphiphilic, and hydrophilic (charged and uncharged) capping agents. We will also address the type of biophysical information that can be obtained.

7.1 Hydrophobic nanoparticles

Recently, we identified that hydrophobic silver nanoparticles covered with the natural fatty acid oleic acid (AgNP-OA), strongly interact with model biomembranes [45]. Using LMs of pure phospholipids and a stratum corneum mimic (SCM) membrane, we characterized the rheology and topography of the resulting films in the presence of the AgNP-OA. We observed a generalized expansion of the experimental isotherms with respect to the ideal ones, with a concomitant diminishing in the compressibility modulus (C_s^{-1}). This behavior clearly reflects an increase in the in-plane elasticity of the films in presence of the AgNP-OA and appears to be augmented in films having more condensed and organized phase states [45].

In previous work, we reported the interaction of hydrophobic MNP-OA, with different phospholipids and mixtures of them [31]. Contrary to the observations made for the AgNP-OA, we detected a selective interaction among the MNP-OA with saturated and unsaturated phospholipids. Moreover, there were interesting differences between the behaviors that both NPs presented when there was a lipid film that offers an LE–LC phase coexistence. While the AgNP-OA showed a partition on LC phases, there was a clear preference of MNP-OA to segregate in LE phases [31]. The unequivocal implication of the *core* composition on the performance that NPs show on the interaction with phospholipids [75], opens a possibility to understand which structural factors are responsible of guiding the final response of NPs towards the biomolecules that represent the principal components of cell membranes. Herein, we propose a brief discussion of the main key factors that regulate the interaction between the aforementioned hydrophobic nanoparticles and biomembranes with phase coexistence. Fig. 6 shows the Langmuir isotherms and BAM micrographs for AgNP-OA and MNP-OA mixed with the phospholipid DPPC (1,2-dipalmitoyl-*sn*-glycero-3-phosphocholine). As evidenced by Fig. 6A, the incorporation of MNP-OA caused a marked reduction of the film area if compared with the ideal isotherm indicating the loss of phospholipids from the interface. The BAM micrographs showed the preferential segregation of MNP-OA in the LE phase. As the monolayer was compressed the NPs caused an impairment in the coalescence of the liquid condensed domains of the DPPC (Fig. 6D) as occurred with the single lipid (Fig. 6C).

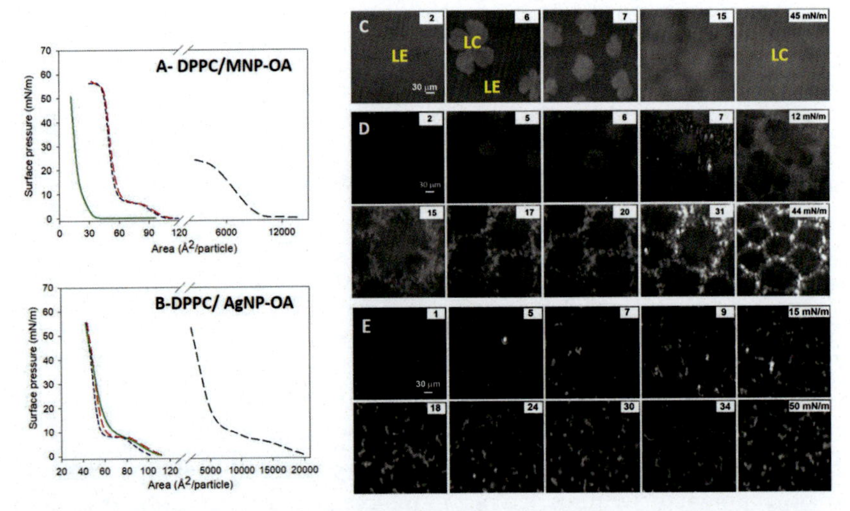

Fig. 6 Surface pressure–area isotherms at the air–water interface for DPPC in presence of MNP-OA (A) and AgNP-OA (B). Pure DPPC (blue, short-dashed line), pure NPs (black, long-dashed line), experimental (green, solid line) and ideal mixed monolayers with $X^{AgNP\text{-}OA} = 0.0005$ and $X^{MNP\text{-}OA} = 0.0003$ (red, long-dashed line). BAM images showing the topography of the films in absence (C) and in presence of MNP-OA (D) or AgNP-OA (E) at the indicated surface pressures. Scale bar 30 μm.

The organizational configuration determined by the magnetic nanoparticles also led to a loss of line tension in the condensed lipidic domains (triskelions). Subsequent surface pressure increments promoted the formation of a long-range hexagonal lattice of LC domains of DPPC surrounded by a net of MNP-OA [31].

For AgNP-OA, in contrast to what was observed for single-phase phospholipids, the interaction with the biphasic DPPC phospholipid showed a less notorious film expansion (Fig. 6B). Herein, almost ideal-like trends were seen at the onset of the monolayer compression isotherm in the gas phase and near the monolayer collapse region. Even though a slight expansion was produced in the LE–LC phase transition region (see the isotherm at ~80–60 $Å^2$/particle); here the most striking fact is that the transition zone itself seems blurred as also evidenced by BAM. This phenomenon indicates that AgNP-OA are able to establish a more cooperative behavior in the interaction with DPPC than MNP-OA. The BAM images enabled us to visualize that AgNP-OA are preferably disposed at the edges of the condensed domains of the phospholipid without causing notorious alterations on the line tension of the lipidic structures (Fig. 6E). AgNP-OA did not display a long-range regular pattern, as observed for MNP-OA, but their segregation on the condensed domains of DPPC along the entire compression seems to cause a decrease in the coalescence of the LC structures, as well as a retardation in the formation of a continuous condensed phase. It is noticeable that at all surface pressures, AgNP-OA tends to interact among them forming nanoparticles groups or clusters at the surface; these clusters of AgNP-OA remained with this organization at all surface pressures without acquiring a uniform distribution on the surface, suggesting a strong interparticle interaction once established.

7.2 Hydrophilic and amphiphilic nanoparticles

Owing to the biocompatibility and the metabolism of natural polysaccharides into nontoxic degradation byproducts, certain biopolymers, such as starch, chitosan and dextran have found promising biomedical applications in surgery, tissue engineering, drug-delivery and oncology [76]. However, only a few works have addressed the implications of NPs functionalized with these polysaccharides when they interact with LMs as will be addressed next.

Dextran-coated IOMNPs (IOMNP–dextran) are currently available in the market to treat iron deficiency in chronic kidney disease [77]. Also, until 2008 IOMNP–dextran were used as MRI (magnetic resonance imaging) contrast agents to detect tumor cells and local lesions in vivo, but this product has been discontinued because of proven side effects [77,78]. The effects induced on pure lipid monolayers of DPPC or DPPG (1,2-dipalmitoyl-*sn*-glycero-3-phospho-(1′-rac-glycerol)) at the air/water interface by IOMNP–dextran dispersed in neutral aqueous subphase has been evaluated by Uehara et al. [71]. Although the presence of pure dextran in the subphase did not exert any changes on the Π–A isotherm of DPPC, when the polymer was bound to the IOMNP surface, a shift of the isotherm to lower molecular areas upon its compression was recorded. This condensation effect was attributed to the adsorption of DPPC molecules on the IOMNP–dextran surface. As proved by SFG spectroscopy, the final organization at high surface pressure can be depicted as lipid molecules with their tails in a well-packed undisturbed fashion at the interface, with DPPC-coated IOMNP–dextran dislocated on top of the condensed monolayer [71]. On the other hand, neither the pure dextran nor the IOMNP–dextran caused any changes on the isotherm of DPPG. Here one needs to consider that the IOMNP–dextran bared negative Z potential in pure water, as stated by the authors, possibly due to –OH groups arising from the iron oxide surface [79,80]. Therefore, the interactions of IOMNP–dextran with anionic DPPG and zwitterionic DPPC were proposed to be mainly driven by electrostatic interactions [71].

Mixed monolayers of starch-coated IOMNPs (IOMNP–starch) and DPPC spread at the air/water interface exhibited a displacement of the Π–A isotherms towards lower molecular areas at high surface pressures with respect to pure DPPC. Therefore, as with IOMNP–dextran, the desorption of DPPC molecules from the interface by IOMNP–starch was also suggested as a plausible explanation [72]. Additionally, BAM images showed the same general tendency of polysaccharide-coated IOMNP to form clusters of NPs visible in the mesoscale which preferentially remain in the LE phase rather than in the LC phase of DPPC regardless of the polarity of the capping This trend was also observed for hydrophobic MNP-OA with DPPC [31]. The observed segregation of IOMNP–starch was indicative of partial or total immiscibility of the particles in the LC phase of DPPC. As lateral compression continued, the IOMNP–starch clusters finally appeared randomly distributed around the LC phase. The capability of IOMNP–starch to reach and remain adsorbed at the DPPC/air interface was validated by the increment of the surface pressure on time after the injection of the NPs underneath a DPPC monolayer in LC state [72].

Biocompatible synthetic polymers have also been used to coat IOMNPs. This is the case of IOMNPs coated with catechol-terminated copolymers of di(ethylene glycol) methyl ether methacrylate (MEO_2MA) and poly(ethylene glycol) methyl ether methacrylate (OEGMA) (IOMNP-MEO_2MA_{90}-*co*-$OEGMA_{10}$) [68]. The authors reported, from adsorption experiments, that IOMNP-MEO_2MA_{90}-*co*-$OEGMA_{10}$ could only penetrate DPPC LMs compressed up to 25 mN/m. This value also matched the equilibrium surface pressure of both IOMNP-MEO_2MA_{90}-*co*-$OEGMA_{10}$, and the block copolymer itself adsorbed at the bare air/water interface. Therefore, the maximum insertion pressure of the coated IOMNPs was controlled by the surface activity of the coating. Accordingly, the interaction of these particles, either spread in a mixture with DPPC or dispersed in the aqueous subphase at whose interface was seeded a DPPC monolayer, caused the shifting of the Π–A isotherms to higher molecular areas compared to pure DPPC. This indicates that the particles are incorporated at the air/lipid interface. As lateral compression progressed, the coated IOMNPs reorganized and remained anchored at the interface favored not only by their intrinsic surface activity, but also due to positive interactions with DPPC head groups. At high lateral pressures above 25 mN/m, the

IOMNPs squeezed out from the interface and the DPPC monolayer regained its original packing density. The BAM images of the mixed monolayers indicated phase separation between the coated IOMNP-MEO_2MA_{90}-*co*-$OEGMA_{10}$ and DPPC at low surface pressures with segregated LE domains in the form of patches. With further compression, the formation of a homogeneous LC lipid phase was inhibited by the segregated IOMNP-MEO_2MA_{90}-*co*-$OEGMA_{10}$. After overpassing the critical surface pressure of 25 mN/m, the desorption of the IOMNP-MEO_2MA_{90}-*co*-$OEGMA_{10}$ from the interface occurred, and an essentially homogeneous LC phase of DPPC was formed. In addition, GIXD results indicated a reduction of the tilt angle of DPPC in contact with the IOMNPs. The increased in-plane packing of the monolayer structure was attributed to changes in the reorientation and/or hydration of the lipid head groups upon interaction with IOMNPs, which ultimately led to free interfacial space to be occupied by the IOMNPs. The results on the interaction of this type of polymer-coated IOMNP with LMs were generalized to include other observations made with polyethylene glycol copolymers [68].

Regarding IOMNPs coated with stabilizing agents that render negative surface charge, only the effect of citrate-coated IOMNPs (IOMNP-citrate) on mixed DOPC/cholesterol/fluorophore-tagged rhodamine B sulfonyl DOPE (Rhod-DOPE) (89:10:1 mol%) LMs has been reported [70]. The Π–A isotherm of the multicomponent monolayer spread on an aqueous dispersion of IOMNP–citrate appeared at higher molecular areas at the whole compression isotherm compared to the lipid monolayer in the absence of the particles. As observed with other examples, this result could indicate either intercalation of IOMNP–citrate into the interface prior or during the compression; or a tilt of the lipid molecules, causing them to occupy larger surface areas. The BAM images of the mixed monolayer showed the coexistence of LE and LC phases from the beginning of the compression, followed by enlargement of the LC domains until a uniform condensed phase was formed. However, with increasing lateral compression, the presence of IOMNP–citrate in the subphase interfered with the LE–LC phase transition of the monolayer, causing changes in the size and shape distributions of the LC domains and inhibiting the formation of a fully condensed continuous film. Although the particle clusters were not detected as in other BAM images, the presence of the particles induced little holes of LE domains with the LC phase in the background, which persisted even at high surface pressures. It is argued that the interaction between the particles and the mixed monolayer altered the cohesion between the lipid molecules, affecting the molecular packing, and preventing the lipid molecules from adopting a more ordered interfacial organization [70].

Following the use of negatively charged NPs, it has been demonstrated that hydrophilic AgNPs functionalized with 4-mercaptobenzoic acid (MBA) have the ability to diffuse and adsorb irreversibly into DMPC preformed Langmuir monolayers. On the other hand, citrate-coated AgNPs having the same core size and similar sectional area, are prone to be excluded from DMPC monolayer upon compression at the air/water interface [57]. In this regard, it was proposed that the permanence of AgNPs–MBA upon their interfacial adsorption and further compression, can be interpreted on the basis of a synergy between electrostatic and hydrophobic interactions. On one side, the AgNPs–MBA are able to excerpt strong electrostatic interactions with the positive ammonium moiety of the polar end of DMPC. On the other side, they can also develop hydrophobic forces between the MBA aromatic groups and the lipid tail protrusions of the phospholipids. AgNPs–MBA can even develop these forces at the phospholipid monolayer defects where the aromatic ring/hydrocarbon chain proximity is favored [57].

Starch and chitosan both derivatized with ethylenediamine have been employed as capping agents of IOMNPs (IOMNP–starch–NH_2 and IOMNP–chitosan–NH_2, respectively) to increase water-dispersibility and surface chemical reactivity of the particles due to amine end groups. The interactions of both modified IOMNPs with DPPC monolayers were evaluated by Piosik et al. [58,59,72]. Independently of the NP capping nature (starch-NH_2 or chitosan-NH_2), the isotherms of the mixed NP/lipid monolayers revealed a shifting to higher molecular areas as the amount of NPs in the mixture increased, accompanied by the vanishing of the characteristic LE–LC phase transition of pure DPPC [81]. The results indicated that, although both types of NPs did not adsorb at the bare air/water interface, they were drawn to the interface even before starting the compression in the presence of the phospholipid monolayer. As a result, the effect of the presence of the NPs at the interface caused expansions of the Π–A isotherms and lower C_s^{-1}, which reflected a disruption of the natural lipid lateral packing density and reorganization upon compression. The effects induced by both types of NPs in the lipid phase transition were confirmed by BAM. Following the same trend of coated-IOMNPs, IOMNP–starch–NH_2 and IOMNP–chitosan–NH_2 preferably segregated in the LE phase rather than in the LC. Nonetheless, the LC domains finally coalesced and continuous films formed with NPs cluster within. This observation was consistent with the higher limiting areas recorded for the compressed mixed NP/DPPC systems. The adsorption experiments indicated that both types of aminated hydrophilic IOMNPs possessed the capability to adsorb onto DPPC monolayers in the LC phase after their introduction into the aqueous subphase and reached higher equilibrium surface pressure than the initial surface pressure value [59,72]. In the case of IOMNP–chitosan–NH_2, the value of the final surface pressure increased with the amount of NPs injected in the DPPC monolayer subphase [58].

Based on the comparative analysis of the presented case studies, we highlight the correlation that appears between the chemical nature of the surface coating of IOMNPs (i.e. hydrophilic neutral or positively-charged), and the effects produced on the surface activity and mesoscale reorganization of phospholipid LMs. To reinforce the idea about the importance of the surface chemical nature of coated-IOMNPs, there is the work by Uehara et al. in which the effect of free poly(diallydimethylammonium chloride) (PDAC) and the polycation-bound to IOMNPs (IOMNP–PDAC) both in water on DPPC or DPPG LMs were studied [71]. Herein, both the free and bound polymer caused the expansion of DPPG monolayers towards larger molecular areas. On the contrary, DPPC molecules were expulsed from the interface by the free and bound polymer, as reflected by the shifting of the Π–A isotherms to the opposite direction (smaller areas). This dependency with the lipid structure was also reported for hydrophobic MNP-OA interacting with saturated and unsaturated phosphocholines [31]. Although SFG spectroscopy revealed that the packing and orientation of both lipids remained unaffected by the presence of IOMNP–PDAC in the subphase, the nature of the polar group of the lipid governed the final picture of the compressed film. On one hand, the negative charge of DPPG allowed the intercalation of positive IOMNP–PDAC at the interface to form stable mixed monolayers. On the other hand, IOMNP–PDAC caused the desorption, and were surrounded by DPPC molecules rendering particle/lipid complexes that translocated on top of the DPPC monolayer. IOMNPs stabilized by aminated polysaccharides or PDAC, which are hydrophilic and positively-charged polymers, induced divergent effects on the properties of DPPC monolayers [64]. Here one needs to bear in mind that, apart from the charge and the chemical nature, the conformational disposition of the capping (topology at NP surface) also plays an important key role in the mechanisms of interaction between coated NPs and lipid LMs. In this sense, it is well known that the topology of polymeric nanoparticles (polymersomes) modulates the endocytic path for NPs cell internalization [82].

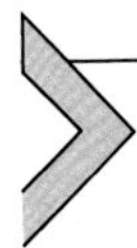

8. Scope, limitations, and perspectives of Langmuir monolayers as membrane models

In the preceding sections we provided actual cases that prove why Langmuir monolayers at the air/water interface have become one of the current preferred models to mimic the outer leaflet of a cell membrane. LMs allows to comes inside several aspects: whether NP will penetrate/adsorb onto the membrane; how the membrane properties will be affected by the NP and vice-versa; and which functional groups would be involved in the interaction. Owing to their versatility, LMs have contributed to understand the correlation between the physiological action and possible toxicity of simple biologically-relevant exogenous molecules (pharmaceuticals, etc.) [83]. However, when it comes to NPs, their intrinsic complexity cannot be fully approached by using simplistic model membranes. This is in part because exact replicates of the bilayer cell composition are unattainable with Langmuir monolayers, and also these bidimensional films do not allow to mimic membrane permeabilization or membrane curvature [20]. Thus, the effects of NPs on planar model membranes does not necessarily correlate with actual observations on living cells. Bearing these limitations in mind, the utility of Langmuir monolayers as a tool to study the mechanisms of interaction of NPs with model biomembranes firstly needs to be expanded to more complex films. Secondly, it has to be conceived as a complementary technique to be used in combination with other approaches with the aim of accurately determining the mechanisms behind the biophysical and physiological action of NPs. This is because basically, there is no single experimental approach that provides a complete picture of the whole processes involved.

As new nanomaterials emerge day to day—leaving aside those specifically designed as nanotherapeutics—, there is a concern that they will eventually find their path into the ecosystem and come into contact with living cells [84]. It needs to be consider that NPs are not inherently benign and they will likely affect biological behaviors at the cellular, subcellular, and protein levels [85]. Hence, it is paramount the implementation of safe-by-design approaches when manufacturing new nanomaterials. There is currently a vast, yet scattered amount of data collected from in vitro and *in cellulo* experiments, as well as *in silico* data modeling of the phenomena taking place at nano-biointerfaces. All of which are expected to translate into quantitative structure–activity relationships and algorithms that serve in the design of efficient and benign NPs [74]. In this context, the application of Langmuir monolayers should be considered as one of the starting points to establish relationships between the effects of the systematic variation of structural properties of NPs on model membranes and their behavior when interacting with actual biomembranes. Multi-component monolayers formed at the air/water interface composed of saturated and unsaturated lipids or lipid mixtures that mimic biological membranes, proteins, polysaccharides, and other biomolecules of interest can be achieved [83]. Additionally, physiological temperature and ionic strength can also be evaluated in the complex films during the experiments, thus resembling actual cell membrane structure and physiology. For example, films formed from cell extracts have been applied [24,86]. The understanding of the way membrane components are formed and structured in the presence of NPs is achievable by means of Langmuir monolayers, yet these processes remain unexplored. Another aspect that is worth exploring is the comparative interaction of NP with model films that mimic abnormal membranes of tumor cells and also, with membrane models of normal cells. The results would help understand and circumvallate enhanced rejection of NPs by tumors and, for example, make cancer drug nanocarriers more effective.

As introduced in previous sections, NPs tend to absorb proteins present in biological media, forming both inner hard and outer soft "protein coronas" that are nonspecifically associated with the NP surface [7,87,88]. As a result, this modified surface actually acts as the real surface that biologically interacts with natural barriers encountered, such as cell membranes. Moreover, the NP surface itself may alter the structure and function of the adsorbed proteins, thus adding extra complexity to the recognition and cellular uptake mechanism of NPs [7,87] The thermodynamic and rheological aspects of lipid monolayers that adsorb/incorporate proteins have been extensively studied [89]. However, experimental evidence of the interaction of NPs with membrane models that incorporate relevant proteins or the interaction of protein-coated NPs with lipid monolayers is still scarce [90]. This knowledge is critical to understand how and why NPs actual responses in biological environments may completely differ from those of the originally designed NPs [74]. As an example of the importance of incorporating proteins in these studies and the need of adding complexity to commonly used single lipid monolayers is the case of hydrophobic AuNPs. It was found that the role of the hydrophobic capping is not innocuous because it can modulate the lipid–NP interactions depending on the type of phospholipids or capping agent. When DPPC/hexadecanethiolate-capped AuNPs mixtures were spread at the air/water interface, an expansion of Π–A isotherms (shifting to higher molecular areas) of the mixed solutions relative to the isotherm of the pure lipid took place. This change indicated incorporation of the capped-AuNPs into the DPPC monolayer. On the contrary, no effects were observed when the same functionalized AuNPs were studied using monolayers of a clinical pulmonary surfactant that contains DPPC as the main lipid component mixed with hydrophobic pulmonary surfactant proteins [91]. Therefore, this case study proves the importance of incorporating proteins in Langmuir monolayer studies as they play key roles in the interactions with NPs.

Discerning specific and nonspecific interactions of NPs with cell membranes and the role of molecular recognition by membrane receptors is also crucial. The interaction of NPs with specific membrane receptors is likely to be enhanced as opposed to the binding of free receptors due to favorable entropic contributions. The attachment of NPs to certain membrane receptors could lead to subsequent protein expression and specific mechanisms of internalization [83]. As a matter of fact, the incubation of nanomaterials with cells results in adsorption of serum proteins on their surface, inducing the entry of nanoparticles into cells by receptor-mediated endocytosis [11,92].

For example, the coating of NPs with opsonins, such as immunoglobulin G (IgG) effectively makes them prone to clearance by phagocytosis, as opposed to surface coating with glycoproteins that were able to bypass phagocytic recognition by macrophages [93]. The coating of NPs with PEG has been used as an important design feature to mimic cell's glycocalyx, and thus preventing opsonization and bypassing phagocyte-mediated cellular barriers [94]. Langmuir monolayers are suitable to mimic recognition sites in order to determine the nature of specific interactions of NP [83].

To conclude, the on-going emergence of novel nanomaterials with promising features demands physiological and potential toxicity testing that ensure safe use in humans and low ecotoxicity. These practices must comply with ethical and cruelty-free principles in alignment with their expected minimal health risk. The Langmuir monolayers as a standard procedure to characterize the interaction of newly developed NPs with model biomembranes advocates to minimize the use of classic animal testing as well as minimize the high cost of production of experimenting with in vivo models.

Acknowledgments

Authors acknowledge the financial support of Consejo Nacional de Investigaciones Científicas y Técnicas (CONICET), Agencia Nacional de Promoción Científica y Tecnológica (ANPCyT), and the Secretaría de Ciencia y Tecnología, Universidad Nacional de Córdoba (SECyT–UNC), Argentina. M.E.V. acknowledges financial support of 'HTMSoft' and 'DELTA' projects number 40003040 and 40008129 by 'Fonds de la Recherche Scientifique' (FNRS). S.D.S and R.V.V. acknowledge CONICET for the financial support.

References

[1] F. Caruso, T. Hyeon, V. Rotello, The unique role of nanoparticles in nanomedicine: imaging, drug delivery and therapyw, Chem. Soc. Rev. 41 (2012) 2885–2911.
[2] R. Mout, D.F. Moyano, S. Rana, V.M. Rotello, Surface functionalization of nanoparticles for nanomedicine, Chem. Soc. Rev. 41 (2012) 2539–2544.
[3] A.E. Nel, L. Mädler, D. Velegol, T. Xia, E.M.V. Hoek, P. Somasundaran, et al., Understanding biophysicochemical interactions at the nano-bio interface, Nat. Mater. 8 (2009) 543–557.
[4] C.J. Kirkpatrick, W. Bonfield, NanoBioInterface: a multidisciplinary challenge, J. R. Soc. Interface 7 (2010) S1–S4.
[5] L. Digiacomo, D. Pozzi, S. Palchetti, A. Zingoni, G. Caracciolo, Impact of the protein corona on nanomaterial immune response and targeting ability, Wiley Interdiscip. Rev.: Nanomed. Nanobiotechnol. 12 (2020) 1–15.
[6] T. Kopac, Protein corona, understanding the nanoparticle–protein interactions and future perspectives: a critical review, Int. J. Biol. Macromol. 169 (2021) 290–301.
[7] R.M. Pearson, V.V. Juettner, S. Hong, Biomolecular corona on nanoparticles: a survey of recent literature and its implications in targeted drug delivery, Front. Chem. 2 (2014) 1–7.
[8] B. Kharazian, N.L. Hadipour, M.R. Ejtehadi, Understanding the nanoparticle-protein corona complexes using computational and experimental methods, Int. J. Biochem. Cell Biol. 75 (2016) 162–174.
[9] M. Lundqvist, C. Augustsson, M. Lilja, K. Lundkvist, B. Dahlbäck, S. Linse, et al., The nanoparticle protein corona formed in human blood or human blood fractions, PLoS One 12 (2017) 1–15.
[10] S. Zhang, H. Gao, G. Bao, Physical principles of nanoparticle cellular endocytosis, ACS Nano 9 (2015) 8655–8671.
[11] G. Canton, I. Battaglia, Endocytosis nanoscale, Chem. Soc. Rev. 41 (2011) 2718–2739.
12] A.M. Farnoud, S. Nazemidashtarjandi, Emerging investigator series: interactions of engineered nanomaterials with the cell plasma membrane; what have we learned from membrane models? Environ. Sci. Nano 6 (2019) 13–40.
[13] K.L. Chen, G.D. Bothun, Nanoparticles meet cell membranes: probing nonspecific interactions using model membranes, Environ. Sci. Technol. 48 (2014) 873–880.

[14] L.A. Clifton, R.A. Campbell, F. Sebastiani, J. Campos-Terán, J.F. Gonzalez-Martinez, S. Björklund, et al., Design and use of model membranes to study biomolecular interactions using complementary surface-sensitive techniques, Adv. Colloid Interface Sci. 277 (2020) 1–22.
[15] E. Rascol, J.M. Devoisselle, J. Chopineau, The relevance of membrane models to understand nanoparticles-cell membrane interactions, Nanoscale 8 (2016) 4780–4798.
[16] G. Brezesinski, H. Möhwald, Langmuir monolayers to study interactions at model membrane surfaces, Adv. Colloid Interface Sci. 100–102 (2003) 563–584.
[17] H.L. Brown, R.E. Brockman, Using monomolecular films to characterize lipid lateral interactions, Methods Mol. Biol. 398 (2007) 41–58.
[18] J.J. Giner-Casares, G. Brezesinski, H. Möhwald, Langmuir monolayers as unique physical models, Curr. Opin. Colloid Interface Sci. 19 (2014) 176–182.
[19] L. Caseli, T.M. Nobre, A.P. Ramos, D.S. Monteiro, M.E.D. Zaniquelli, The role of Langmuir monolayers to understand biological events, ACS Symp. Ser. 1215 (2015) 65–88.
[20] J. Rascol, E. Devoisselle, J.-M. Chopineau, The relevance of membrane models to understand nanoparticles-cell membrane interactions, Nanoscale 8 (2016) 4780–4798.
[21] R. Pignatello, T. Musumeci, L. Basile, C. Carbone, G. Puglisi, Biomembrane models and drug-biomembrane interaction studies: involvement in drug design and development, J. Pharm. Bioallied Sci. 3 (2011) 4–14.
[22] H.L. Brockman, Lipid monolayers: why use half a membrane to characterize protein-membrane interactions? Curr. Opin. Struct. Biol. 9 (1999) 438–443.
[23] N. Wilke, Lipid monolayers at the air-water interface. A tool for understanding electrostatic interactions and rheology in biomembranes, Advances in Planar Lipid Bilayers and Liposomes, Elsevier B.V., 2014, pp. 51–81.
[24] M. Rojewska, W. Smułek, E. Kaczorek, K. Prochaska, Langmuir monolayer techniques for the investigation of model bacterial membranes and antibiotic biodegradation mechanisms, Membranes 11 (2021) 1–20.
[25] L. Song, B.B. Xu, Q. Cheng, X. Wang, X. Luo, X. Chen, et al., Instant interfacial self-assembly for homogeneous nanoparticle monolayer enabled conformal 'lift-on' thin film technology, Sci. Adv. 7 (2021) 1–9.
[26] J. Flesch, M. Kappen, C. Drees, C. You, J. Piehler, Self-assembly of robust gold nanoparticle monolayer architectures for quantitative protein interaction analysis by LSPR spectroscopy, Anal. Bioanal. Chem. 412 (2020) 3413–3422.
[27] G. Yang, D.T. Hallinan, Gold nanoparticle monolayers from sequential interfacial ligand exchange and migration in a three-phase system, Sci. Rep. 6 (1) (2016) 17.
[28] D. Vollhardt, Brewster angle microscopy: a preferential method for mesoscopic characterization of monolayers at the air/water interface, Curr. Opin. Colloid Interface Sci. 19 (2014) 183–197.
[29] G. Gaines, Insoluble Monolayers at Liquid-Gas Interfaces, first ed., Interscience Publishers, New York, 1966.
[30] E.K. Davies, J.T. Rideal, Interfacial Phenomena, fourth ed., Academic Press Inc., New York, 1961.
[31] T.J. Matshaya, A.E. Lanterna, A.M. Granados, R.W.M. Krause, B. Maggio, R.V. Vico, Distinctive interactions of oleic acid covered magnetic nanoparticles with saturated and unsaturated phospholipids in langmuir monolayers, Langmuir 30 (2014) 5888–5896.
[32] M. Jurak, K. Szafran, P. Cea, S. Martín, Analysis of molecular interactions between components in phospholipid-immunosuppressant-antioxidant mixed Langmuir films, Langmuir 37 (2021) 5601–5616.

[33] F. Dupuy, B. Maggio, The hydrophobic mismatch determines the miscibility of ceramides in lipid monolayers, Chem. Phys. Lipids 165 (2012) 615–629.
[34] S. Ali, J.M. Smaby, H.L. Brockman, R.E. Brown, Cholesterol's interfacial interactions with galactosylceramides, Biochemistry 33 (1994) 2900–2906.
[35] J.J. Pinzón Barrantes, B. Maggio, R.H. De Rossi, R.V. Vico, Cavity orientation regulated by mixture composition and clustering of amphiphilic cyclodextrins in phospholipid monolayers, J. Phys. Chem. B 121 (2017) 4482–4491.
[36] D. Marsh, Lateral pressure in membranes, Biochim. Biophys. Acta Rev. Biomembr. 1286 (1996) 183–223.
[37] M. Mottola, R.V. Vico, M.E. Villanueva, M.L. Fanani, Alkyl esters of l-ascorbic acid: stability, surface behaviour and interaction with phospholipid monolayers, J. Colloid Interface Sci. 457 (2015) 232–242.
[38] M. Zulueta Diaz, Y. Mottola, M.L. Vico, R. Wilke, N. Fanani, The rheological properties of lipid monolayers modulate the incorporation of L-ascorbic acid alkyl esters, Langmuir 32 (2016) 587–595.
[39] A. Hädicke, A. Blume, Interactions of pluronic block copolymers with lipid monolayers studied by epi-fluorescence microscopy and by adsorption experiments, J. Colloid Interface Sci. 407 (2013) 327–338.
[40] A. Maestro, E. Guzmán, F. Ortega, R.G. Rubio, Contact angle of micro- and nanoparticles at fluid interfaces, Curr. Opin. Colloid Interface Sci. 19 (2014) 355–367.
[41] M. Zeng, J. Mi, C. Zhong, Wetting behavior of spherical nanoparticles at a vapor-liquid interface: a density functional theory study, Phys. Chem. Chem. Phys. 13 (2011) 3932–3941.
[42] V. Garbin, J.C. Crocker, K.J. Stebe, Nanoparticles at fluid interfaces: exploiting capping ligands to control adsorption, stability and dynamics, J. Colloid Interface Sci. 387 (2012) 1–11.
[43] X. Hua, J. Frechette, M.A. Bevan, Nanoparticle adsorption dynamics at fluid interfaces, Soft Matter 14 (2018) 3818–3828.
[44] J.Y. Kim, S. Raja, F. Stellacci, Evolution of Langmuir film of nanoparticles through successive compression cycles, Small 7 (2011) 2526–2532.
[45] M.E. Villanueva, A.E. Lanterna, R.V. Vico, Hydrophobic silver nanoparticles interacting with phospholipids and stratum corneum mimic membranes in Langmuir monolayers, J. Colloid Interface Sci. 543 (2019) 247–255.
[46] F. Stellacci, C.A. Bauer, T. Meyer-Friedrichsen, W. Wenseleers, V. Alain, S.M. Kuebler, et al., Laser and electron-beam induced growth of nanoparticles for 2D and 3D metal patterning, Adv. Mater. 14 (2002) 194–198.
[47] F. Bresme, M. Oettel, Nanoparticles fluid interfaces, J. Phys. Condens. Matter 19 (2007) 1–33.
[48] D. Borchman, E.N. Harris, S.S. Pierangeli, O.P. Lamba, Interactions and molecular structure of cardiolipin and β2-glycoprotein 1 (β2-GP1), Clin. Exp. Immunol. 102 (1995) 373–378.
[49] W.M. Gelbart, R.P. Sear, J.R. Heath, S. Chaney, Array formation in nano-colloids: theory and experiment in 2D, Faraday Discuss. 122 (1999) 299–307.
[50] G.V. Bossa, K. Bohinc, M.A. Brown, S. May, Dipole moment of a charged particle trapped at the air-water interface, J. Phys. Chem. B 120 (2016) 6278–6285.
[51] P. Pieranski, Two-dimensional interfacial colloidal crystals, Trans. Faraday Soc. 45 (1980) 569–572.
[52] R. Aveyard, J.H. Clint, D. Nees, V.N. Paunov, Compression and structure of monolayers of charged latex particles at air/water and octane/water interfaces, Langmuir 16 (2000) 1969–1979.
[53] C. Binns, Medical applications of magnetic nanoparticles, Frontiers of Nanoscience, Elsevier Ltd., 2014, pp. 217–258.

[54] S. Lefebure, C. Ménager, V. Cabuil, M. Assenheimer, F. Gallet, C. Flament, Langmuir monolayers of monodispersed magnetic nanoparticles coated with a surfactant, J. Phys. Chem. B 102 (1998) 2733–2738.
[55] M.I. Viseu, A.M.G. da Silva, S.M.B. Costa, Reorganization and desorption of catanionic monolayers. Kinetics of π-t and A-t relaxation, Langmuir 17 (2001) 1529–1537.
[56] M.E. Villanueva, S.R. Salinas, R.V. Vico, I.D. Bianco, Surface characterization and interfacial activity of chitinase chi18-5 against chitosan in langmuir monolayers, Colloids Surf. B Biointerfaces 227 (2023) 113337.
[57] J.V. Maya Girón, R.V. Vico, B. Maggio, E. Zelaya, A. Rubert, G. Benítez, et al., Role of the capping agent in the interaction of hydrophilic Ag nanoparticles with DMPC as a model biomembrane, Environ. Sci. Nano 3 (2016) 462–472.
[58] E. Piosik, M. Ziegler-Borowska, D. Chełminiak-Dudkiewicz, T. Martyński, Effect of aminated chitosan-coated Fe_3O_4 nanoparticles with applicational potential in nanomedicine on DPPG, DSPC, and POPC Langmuir monolayers as cell membrane models, Int. J. Mol. Sci. 22 (2021) 1–16.
[59] E. Piosik, P. Klimczak, M. Ziegler-Borowska, D. Chełminiak-Dudkiewicz, T. Martyński, A detailed investigation on interactions between magnetite nanoparticles functionalized with aminated chitosan and a cell model membrane, Mater. Sci. Eng. C. 109 (2020) 110616.
[60] T. Gehring, T.M. Fischer, Diffusion of nanoparticles at an air/water interface is not invariant under a reversal of the particle charge, J. Phys. Chem. C 115 (2011) 23677–23681.
[61] J.H. Fendler, Nanoparticles at air/water interfaces, Curr. Opin. Colloid Interface Sci. 1 (1996) 202–207.
[62] D. Ahmad, I. van den Boogaert, J. Miller, R. Presswell, H. Jouhara, Hydrophilic and hydrophobic materials and their applications, Energy Sources Part A: Recovery Util. Environ. Eff. 40 (2018) 2686–2725.
[63] C. Stefaniu, M. Chanana, D. Wang, D.V. Novikov, G. Brezesinski, H. Möhwald, Langmuir and Gibbs magnetite NP layers at the air/water interface, Langmuir 27 (2011) 1192–1199.
[64] Y.J. Shen, Y.L. Lee, Y.M. Yang, Monolayer behavior and langmuir-blodgett manipulation of CdS quantum dots, J. Phys. Chem. B 110 (2006) 9556–9564.
[65] X. Ji, C. Wang, J. Xu, J. Zheng, K.M. Gattás-Asfura, R.M. Leblanc, Surface chemistry studies of (CdSe)ZnS quantum dots at the air-water interface, Langmuir 21 (2005) 5377–5382.
[66] M. Pauly, B.P. Pichon, A. Demortière, J. Delahaye, C. Leuvrey, G. Pourroy, et al., Large 2D monolayer assemblies of iron oxide nanocrystals by the Langmuir-Blodgett technique, Superlattices Microstruct. 46 (2009) 195–204.
[67] S.S. You, R. Rashkov, P. Kanjanaboos, I. Calderon, M. Meron, H.M. Jaeger, et al., Comparison of the mechanical properties of self-assembled langmuir monolayers of nanoparticles and phospholipids, Langmuir 29 (2013) 11751–11757.
[68] C. Stefaniu, G. Brezesinski, H. Möhwald, Polymer-capped magnetite nanoparticles change the 2D structure of DPPC model membranes, Soft Matter 8 (2012) 7952–7959.
[69] G.D. Bothun, N. Ganji, I.A. Khan, A. Xi, C. Bobba, Anionic and cationic silver nanoparticle binding restructures net-anionic PC/PG monolayers with saturated or unsaturated lipids, Langmuir 33 (2017) 353–360.
[70] D. Nieciecka, A. Królikowska, K. Kijewska, G.J. Blanchard, P. Krysinski, Hydrophilic iron oxide nanoparticles probe the organization of biomimetic layers: electrochemical and spectroscopic evidence, Electrochim. Acta 209 (2016) 671–681.

[71] T.M. Uehara, V.S. Marangoni, N. Pasquale, P.B. Miranda, K.B. Lee, V. Zucolotto, A detailed investigation on the interactions between magnetic nanoparticles and cell membrane models, ACS Appl. Mater. Interfaces 5 (2013) 13063–13068.
[72] E. Piosik, A. Zaryczniak, K. Mylkie, M. Ziegler-borowska, Probing of interactions of magnetite nanoparticles coated with native and aminated starch with a DPPC model membrane, Int. J. Mol. Sci. 22 (2021) 5939.
[73] G.B. Soriano, R. da Silva Oliveira, F.F. Camilo, L. Caseli, Interaction of non-aqueous dispersions of silver nanoparticles with cellular membrane models, J. Colloid Interface Sci. 496 (2017) 111–117.
[74] H. Meng, W. Leong, K.W. Leong, C. Chen, Y. Zhao, Walking the line: the fate of nanomaterials at biological barriers, Biomaterials 174 (2018) 41–53.
[75] J. Jeevanandam, A. Barhoum, Y.S. Chan, A. Dufresne, M.K. Danquah, Review on nanoparticles and nanostructured materials: history, sources, toxicity and regulations, Beilstein J. Nanotechnol. 9 (2018) 1050–1074.
[76] M. Grumezescu, L. Li, Multi-bit biomemristic behavior for neutral multi-bit biomemristic behavior for neutral polysaccharide dextran blended with chitosan, Nanomaterials 12 (2022) 1072.
[77] A.A. Halwani, Development of pharmaceutical nanomedicines: from the bench to the market, Pharmaceutics 14 (1) (2022) 21.
[78] A.C. Anselmo, S. Mitragotri, A review of clinical translation of inorganic nanoparticles, AAPS J. 17 (2015) 1041–1054.
[79] K.S. Sharma, R.S. Ningthoujam, A.K. Dubey, A. Chattopadhyay, S. Phapale, R.R. Juluri, et al., Synthesis and characterization of monodispersed water dispersible Fe_3O_4 nanoparticles and in vitro studies on human breast carcinoma cell line under hyperthermia condition, Sci. Rep. 8 (2018) 1–11.
[80] K. Tao, S. Song, J. Ding, H. Dou, K. Sun, Carbonyl groups anchoring for the water dispersibility of magnetite nanoparticles, Colloid Polym. Sci. 289 (2011) 361–369.
[81] H. Möhwald, Phospholipid monolayers, in: R. Physics, Lipowsky, E. Sackmann (Eds.), Handbook of Biological, Elsevier Science B.V., 1995, pp. 161–211.
[82] A. Akinc, G. Battaglia, Exploting endocytosis for nanomedicines, Cold Spring Harb. Perspect. Biol. 5 (2013) 1–24.
[83] O.N. Oliveira Jr., L. Caseli, K. Ariga, The past and the future of Langmuir and Langmuir-Blodgett Films, Chem. Rev. 122 (2022) 6459–6513.
[84] J. Cancino, T.M. Nobre, O.N. Oliveira, S.A.S. Machado, V. Zucolotto, A new strategy to investigate the toxicity of nanomaterials using Langmuir monolayers as membrane models, Nanotoxicology 7 (2013) 61–70.
[85] J. Lin, L. Miao, G. Zhong, C.H. Lin, R. Dargazangy, A. Alexander-Katz, Understanding the synergistic effect of physicochemical properties of nanoparticles and their cellular entry pathways, Commun. Biol. 3 (1) (2020) 10.
[86] C.M. Rosetti, B. Maggio, Protein-induced surface structuring in myelin membrane monolayers, Biophys. J. 93 (2007) 4254–4267.
[87] R. Bilardo, F. Traldi, A. Vdovchenko, M. Resmini, Influence of surface chemistry and morphology of nanoparticles on protein corona formation, WIREs Nanomed. Nanobiotechnol. 14 (2022) 1–22.
[88] J. Ren, N. Andrikopoulos, K. Velonia, H. Tang, R. Cai, F. Ding, et al., Chemical and biophysical signatures of the protein corona in nanomedicine, J. Am. Chem. Soc. 144 (2022) 9184–9205.
[89] M. Elderdfi, A.F. Sikorski, Langmuir-monolayer methodologies for characterizing protein-lipid interactions, Chem. Phys. Lipids 212 (2018) 61–72.
[90] N. Ganji, G.D. Bothun, Albumin protein coronas render nanoparticles surface active: consonant interactions at air-water and at lipid monolayer interfaces, Environ. Sci. Nano 8 (2021) 160–173.

[91] S. Tatur, A. Badia, Influence of hydrophobic alkylated gold nanoparticles on the phase behavior of monolayers of DPPC and clinical lung surfactant, Langmuir 28 (2012) 628–639.
[92] J.J. Rennick, A.P.R. Johnston, R.G. Parton, Key principles and methods for studying the endocytosis of biological and nanoparticle therapeutics, Nat. Nanotechnol. 16 (2021) 266–276.
[93] S. Behzadi, B. Serpooshan, W. Tao, M.A. Hamaly, M.Y. Alkawareek, E.C. Dreaden, et al., Cellular uptake of nanoparticles: journey inside the cell, Chem. Soc. Rev. 46 (2017) 4218–4244.
[94] S.Y. Fam, C.F. Chee, C.Y. Yong, K.L. Ho, A.R. Mariatulqabtiah, W.S. Tan, Stealth coating of nanoparticles in drug-delivery systems, Nanomaterials 10 (2020) 1–18.

Theoretical description of particle sedimentation in blood considering hematocrit: A 2nd generation mathematical model

Maxence Berry[a,b] and Veronika Kralj-Iglič[a,*]
[a]University of Ljubljana, Faculty of Health Sciences, Laboratory of Clinical Biophysics, Ljubljana, Slovenia
[b]University of Poitiers, College of Fundamental and Applied Science, Poitiers, France
*Corresponding author. e-mail address: veronika.kralj-iglic@zf.uni-lj.si

Contents

Abstract

Preparation of therapeutic and diagnostic material from blood requires processing to separate the required materials. As blood is a dynamic fluid, processing may affect the contents of the sample, depending on the intrinsic parameters of blood and its constituents, as well as of the extrinsic parameters (temperature and setting of the equipment). The protocol is key in obtaining samples of required quality and quantity and mathematical modeling is a convenient tool in search for the optimal procedure. Here we present the up-graded mathematical model for preparation of platelet and extracellular vesicles rich plasma (PVRP) by centrifugation of blood with maximal recovery of platelets that takes into account the effect of hematocrit. The model indicates that optimal time and optimal volume of plasma decrease with increasing

Advances in Biomembranes and Lipid Self-Assembly, Volume 38
ISSN 2451-9634, https://doi.org/10.1016/bs.abl.2023.08.002

effectivity of plasma counter-current to erythrocyte sedimentation, increasing hematocrit and increasing amount of sample in the tube, and increases with increasing extension of centrifuge rotor. We have compared the measured and the calculated volume of plasma at optimized centrifugation time and estimated the parameter ε describing the counter-current of plasma for 6 individual samples from two blood donors. We assessed ε for each sample individually. Next, the model must be validated on a larger cohort of different blood samples.

1. Introduction

Plasma can be considered as a platform for cells and different types of particles that are of vital importance within the organism. In practice, the term refers to the material after removing a majority of erythrocytes which are the most abundant type of particles in blood. As blood is relatively easily obtained by venipuncture and erythrocytes can be largely removed by formation of clot (in serum) or sedimentation (in plasma) this material has been extensively studied as well as used in clinics [1–6]. Plasma was reported to have healing properties in different fields of medicine (reviewed in [7]). It was suggested that the healing effect may be due to platelets, therefrom derived growth factors as well as of leukocytes that exert immunomodulatory actions of the immune system (reviewed in [6]). However, the healing mechanisms are not completely understood which limits the success of various plasma-based treatments. It is not yet clear which components of plasma have beneficial effects on particular mechanisms and how these effects can be stimulated.

Recently, extracellular vesicles (EVs) have become an important subject of study due to their mediating role in intercellular communication. EVs are within nano-dimensions (ranging from 20 nm to 1000 nm) and were found in isolates of different body liquids from which blood was among the most studied ones [8–11]. EVs were found to transport cytosolic and membrane proteins and nucleic acids [8] and are considered as important players in the pathophysiology of diseases. Also they are considered as platforms and effectors in novel diagnostic and therapeutic possibilities [9]. In order to emphasize platelets and EVs in processed blood, the term "platelet and extracellular vesicle-rich plasma" (PVRP) has been suggested [2]. In this context, PVRP is considered as the fraction of blood from which erythrocytes were removed while platelets and EV concentrations are comparable or higher with respect to blood.

It is therefore the aim of blood processing to yield plasma with highest possible concentration of platelets and extracellular vesicles [11]. The procedure consists of blood sampling into tubes with a proper amount of anticoagulant and separation of different types of particles. There are many different protocols and there is no standardization which reflects poor understanding of underlying mechanisms.

Centrifugation is a simple and low cost method for plasma preparation. Blood is centrifuged to sediment erythrocytes while plasma constitutes the supernatant of the sample. The acquired plasma may be used already as prepared by a single centrifugation—or—it could be further processed to increase the density of the platelets and EVs [7]. The platelets can be activated endogenously or by stimulation with calcium mixtures before application [7]. The preparation is applied to the targeted area so that the healing substances are delivered to places where blood would otherwise rarely go on its own [7].

Autologous plasma is being applied in different fields of human and veterinary medicine, therefore the need to better understand and master the preparation exists to fully explore its healing potentials. The contents of PVRP depend on the processing methods, i.e. the setting of the centrifuge and external parameters such as temperature. Mathematical modeling can be of aid in interpretation of data and prediction of optimal preparation settings [12,13]. During PVRP preparation, particles in blood undergo a complex motion and are subjected to mutual interactions. However, a useful and transparent model tends to outline the key features that must be taken into account to predict the behavior of the system and get the preparation with preferred properties.

In this work we consider a single centrifugation preparation by a protocol for individualized plasma. A 1st generation simple transparent mathematical model of erythrocyte and platelet motion during centrifugation for PVRP preparation has previously been developed [11]. Within this model, erythrocytes, platelets and EVs in the upper part of the tube are moving according to the centrifugal force, buoyancy and resistance of plasma. Just above the bottom of the tube, erythrocytes are closely packed and plasma that carries platelets and EVs is there moving in the opposite direction. At given centripetal acceleration of the centrifuge rotor, a time was predicted to yield the highest content of platelets in formed plasma for an individual sample [11]. here we tested this model on 6 blood samples from 2 human donors. Furthermore, to improve the theory, we have refined the model by introduction of the effect of hematocrit on optimal time and length of acquired plasma.

2. Theory

2.1 Model of sedimentation

Blood is described as composed of liquid medium with given viscosity η and density ρ in which different particles (blood cells and small particles such as extracellular vesicles, lipid droplets and lipoprotein complexes) move under the effect of relevant forces. Due to different sizes, shapes and surface properties the particles of different types move with different velocities and also in different directions. This is schematically depicted in Fig. 1A. It is envisaged that erythrocytes which are the most numerous and relatively large (Fig. 1B) move within the stripped region towards the

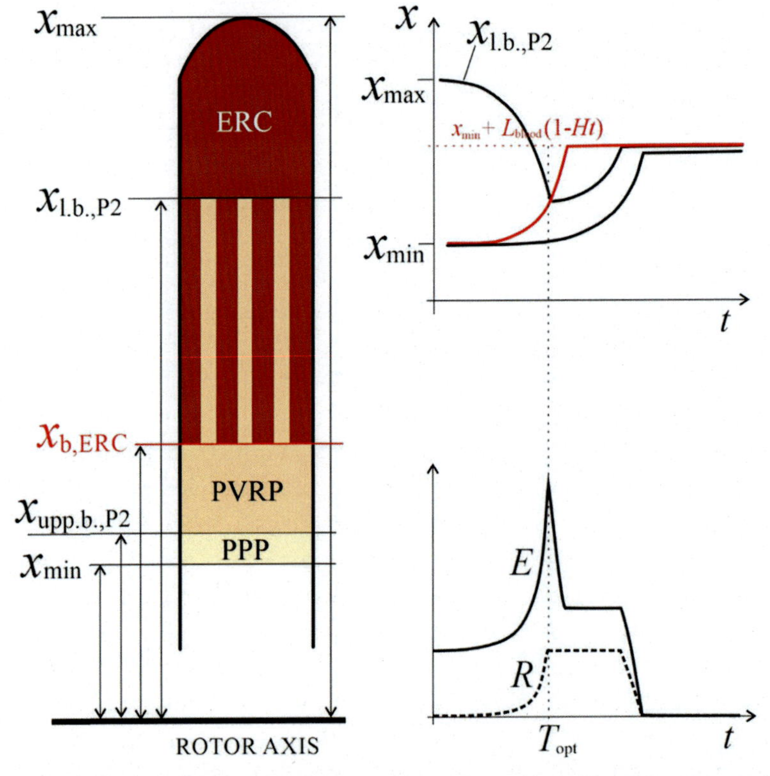

Fig. 1 Model of separation of blood components by centrifugation. P2 represent particles (mostly platelets) within defined flow cytometric gates, E is efficiency of P2 particles yield (ratio between the concentration of P2 particles in plasma and their concentration in blood) and R is recovery of platelets (the number of P2 particles in preparation divided by the number of P2 particles in blood).

bottom of the tube leaving behind plasma with platelets and smaller particles (PPP and PVRP, Fig. 1B).

In the erythrocyte-free region, platelets in PVRP also move towards the bottom of the tube leaving behind platelet poor plasma (PPP, Fig. 1B). When the erythrocytes reach the bottom of the tube their motion is hindered and they closely pack piling up in a region above the bottom of the tube (marked ERC in Fig. 1B). By this, the plasma with smaller particles is squeezed towards the top of the tube. The counter-flow of plasma carrying platelets and smaller particles diminishes the velocity of erythrocytes. Furthermore, the sample is subjected to shear stresses due to a profile of velocity of the particles with respect to the distance from the tube walls. To assess the key characteristics of this complex motion, some simplifications were introduced in the mathematical model [11] which is briefly described below. The boundary effects of the tube were neglected.

The model applies to the movement of particles in the region where accumulation did not yet take place. As swinging rotor was used, we considered that the tube with the sample attains a horizontal position during the centrifugation (Fig. 2). It is imagined that the sample is divided into thin slices with thickness dx (Fig. 2) whereas the concentration and the speed of the particles are constant within the slice.

2.2 Mathematical model

Upon centrifugal acceleration, the particles move from a chosen slice (denoted by index i in Fig. 2) into the one which is closer to the bottom of

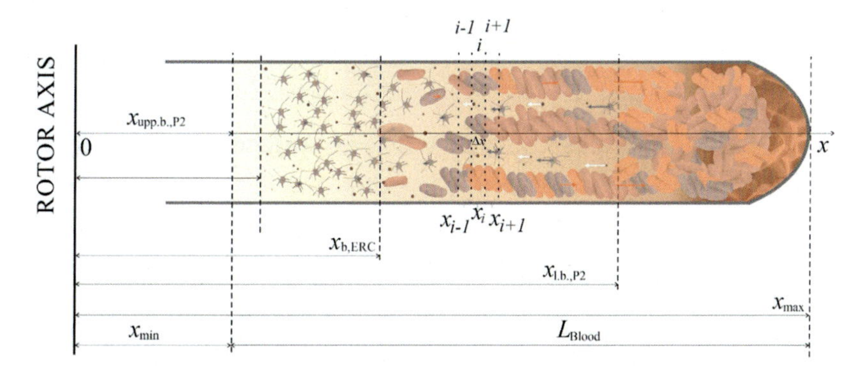

Fig. 2 Scheme of separation of blood constituents during centrifugation in a swinging rotor. *Adapted from D. Božič, D. Vozel, M. Hočevar, M. Jeran, Z. Jan, M. Pajnič, et al., Enrichment of plasma in platelets and extracellular vesicles by the counterflow to erythrocyte settling, Platelets, 33 (2022) 592–602.*

the tube (denoted by index $i+1$ in Fig. 2). At the same time the slice receives the particles from the slice above—the one closer to the rotor axis ($i-1$, Fig. 2). The change of the number of particles in the ith slice N_i with time t is

$$\mathrm{d}N_i/\mathrm{d}t = c_{i-1}\, Su_{i-1}(t) - c_i\, Su_{i.}(t), \tag{1}$$

where c_i is the number density (concentration) of particles, S is the cross section of the tube and u_i is the velocity of the particles in the ith slice at the time t. The cross section of the tube is considered to be constant.

Following [11] it is assumed that the concentration and the velocity change only slightly from one slice to another, so that the expansion can be used to approximate the changes.

$$c_i - c_{i-1} = \mathrm{d}c_i/\mathrm{d}x \quad \Delta x, \tag{2}$$

$$u_i - u_{i-1} = \mathrm{d}u_i/\mathrm{d}x \;\; \Delta x. \tag{3}$$

Neglecting the terms quadratic in Δx and performing the limit $\Delta x \rightarrow \mathrm{d}x$ yields after rearrangement of Eqs. (1)–(3) a differential equation

$$\begin{aligned} &\mathrm{d}c(x, t)/\mathrm{d}t \\ &\quad = -(\mathrm{d}c(x, t)/\mathrm{d}x \;\; u(x, t) + \mathrm{d}u(x, t)/\mathrm{d}x \;\; c(x, t)). \end{aligned} \tag{4}$$

It is assumed that the velocity of the particles is proportional to the centripetal acceleration which is expressed by a multiplicity (X) of the Earth gravity constant g

$$u(x) = \omega X, \tag{5}$$

where ω is a constant characteristic for the sample. It follows from Eqs. (4) and (5) that

$$\mathrm{d}c/\mathrm{d}t = -\omega X \mathrm{d}(cx)/\mathrm{d}x. \tag{6}$$

By using dimensionless quantities: γ—concentration divided by its initial value $c_{\mathrm{ERC},0}$

$$\gamma_{\mathrm{ERC}}(x, t) = c_{\mathrm{ERC}}(x, t)/c_{\mathrm{ERC},0}. \tag{7}$$

ξ—distance divided by the maximal length from rotor axis to the bottom of the tube, $x_{\max}$

$$\xi = x/x_{\max}. \tag{8}$$

and relative sedimentation time τ_{p}—time multiplied by ωX

$$\tau = \omega X \;\; t. \tag{9}$$

Eq. (6) is transformed into

$$d\gamma_p(\xi, \tau)/d\tau = -d(\gamma(\xi, \tau)\xi)/d\xi. \tag{10}$$

An ansatz solution is

$$\gamma(\xi, \tau_p) = C(-\tau) + D \exp(-\tau/2)\xi^{-1/2}, \tag{11}$$

where C and D are constants. As we considered the concentration of particles within plasma departments constant [11], we chose $D = 0$.

The initial condition

$$\gamma_p(0) = 1 \tag{12}$$

implied that $C = 1$ and the relative number density of particles is therefore

$$\gamma(\xi. \tau) = \exp(-\tau). \tag{13}$$

The above model does not apply to the region of the forming sediment where the motion of particles is hindered by direct interactions between each other and with the tube walls.

In experiment we are focusing on concentration of platelets in plasma that is formed above the boundary of the erythrocyte-devoid region ($\xi_{b.,ERC}$). Following Eqs. (10) and (13), the boundary of erythrocytes moves according to

$$d\xi/d\tau = \xi, \tag{14}$$

with the solution

$$\xi_{b.,ERC} = \xi_{min} \exp(\tau). \tag{15}$$

Here $\xi_{min} = x_{min}/x_{max}$ is the dimensionless distance of the surface level of the sample from the rotor axis. A complex motion of particles is assumed to take place in the region below $\xi_{b.,ERC}$ and above $\xi_{l.b.,P2}$ (schematically indicated as striped region in Fig. 1B). It could be envisaged that the removal of plasma from the region of packed erythrocytes were the most efficient if erythrocytes packed into channels that would facilitate the flow of plasma upwards. Erythrocytes are prone to form rouleaux, therefore such configuration is likely to take place. As the plasma is squeezed up by the erythrocytes, we assume that the velocity of the lower bound of the plasma $\xi_{l.b.,P2}$ is opposite in direction while proportional in magnitude to the velocity of erythrocytes,

$$\xi_{l.b.,P2} = \xi_{max} \exp(-(1 + \varepsilon Ht)\tau), \tag{16}$$

where ε is an adjustable parameter and Ht is the hematocrit (ratio between the volume of erythrocytes and volume of blood). We assumed that the number density of erythrocytes would make a difference to the counter-flow of plasma. The expressions (15) and (16) describe the relative change of position and velocity of the boundaries of the particles during centrifugation.

The quantity of interest in clinical application is the volume of plasma acquired. Since processing takes place in tubes with constant cross section S, the volume of plasma $V = SL$ is here represented by the length of the plasma L. In the model, normalized length of the plasma is obtained by the difference between the normalized boundary of erythrocytes [Eq. (15)] and ξ_{min} (Fig. 1B)

$$l = \xi_{min} \exp(\tau) - \xi_{min}. \tag{17}$$

The optimal centrifuge setting is assumed to meet the condition that the lower bound of the plasma meets the bound of the erythrocytes. In this way all the platelets and small particles that are in plasma would be recovered above the erythrocyte boundary and expressions (15) and (16) would equalize to get

$$\tau_{opt} = \ln(\xi_{max}/\xi_{min})/(2 + \varepsilon\, \mathrm{Ht}) \tag{18}$$

and

$$t_{opt} = \ln(\xi_{max}/\xi_{min})/(\omega X\, (2 + \varepsilon\, \mathrm{Ht})) \tag{19}$$

The corresponding optimal length of the plasma is obtained by inserting Eq. (18) into Eq. (17).

$$L_{opt} = \xi_{min} \exp(\tau_{opt}) - \xi_{min}. \tag{20}$$

Using the definition of normalized length [Eq. (8)] we get after some calculation

$$L_{opt} = x_{min}\left((x_{max}/x_{min})^{(1/(2+\varepsilon\, \mathrm{Ht}))} - 1\right). \tag{21}$$

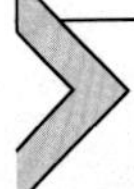

3. Experimental methods

3.1 Sampling of blood

Blood was donated by two of the authors (a female aged 64 and a male aged 22, with no record of disease). Collection was established in the morning after fasting for a minimum of 12 h overnight. A G21 needle (Microlance,

Becton Dickinson, Franklin Lakes. NJ, USA) and 2.7 mL evacuated tubes with trisodium citrate (BD Vacutainers, 367714A, Becton Dickinson, Franklin Lakes, NJ, USA) were used. Blood was processed fresh within 1 h of sampling. While waiting to be centrifuged, the samples were gently mixed on a carousel at room temperature.

3.2 Preparation of plasma

Blood was centrifuged in the tubes in which it was sampled (BD Vacutainers, 367714A, Becton Dickinson, Franklin Lakes, NJ, USA) at 18 °C and 300*g* in the Centric 400R centrifuge (Domel, Železniki, Slovenia) with rotor RS4/100. To test the velocity of the movement of the erythrocyte boundary, blood was first centrifuged for 5 min. The length of the acquired plasma L_{test} and the distances x_{max} and x_{min} were measured by a ruler and τ_{test} was estimated by using Eq. (19),

$$\tau_{test} = \ln(1 + L_{test}/x_{min}). \quad (22)$$

The parameter ω for the individual blood was calculated by using Eqs. (9) and (22),

$$\omega = \ln(1 + L_{test}/x_{min})/(Xt_{test}). \quad (23)$$

The optimal centrifugation time was calculated by using Eqs. (9), (18) and (22).

$$t_{opt} = \ln(x_{max}/x_{min})/(\omega X(2 + \varepsilon\,\mathrm{Ht})). \quad (24)$$

We took $\varepsilon = 0$. Then, blood was centrifuged at 18 °C and 300*g* ($X = 300$) for t_{opt}.

3.3 Determination of hematocrit

Blood was centrifuged at 18 °C and 2000*g* ($X = 2000$) for 15 min. The length of plasma L_{Ht} and the length of the whole sample L_{blood} were measured. Hematocrit was obtained as

$$\mathrm{Ht} = 1 - L_{Ht}/L_{blood}. \quad (25)$$

3.4 Determination of the effectivity of counter-current of plasma ϵ from the experimentally obtained length of plasma *L*

To test the 2nd generation mathematical model, ϵ was adjusted to obtain the length of plasma observed in experiment. ϵ was expressed from Eq. (18) and time *t* was calculated by using Eq. (9) to yield

$$\in = 1/\mathrm{Ht} * (\ln(x_{max}/x_{min})/\ln(1 + L/x_{min}) - 2). \quad (26)$$

4. Results

Table 1 shows estimated optimal time and length of plasma for hypothetical samples with variable parameters ε (representing the effectivity of counter-current of plasma), Ht (the volume of erythrocytes with respect to volume of blood), x_{min} (the amount of sample in the tube) and x_{max} (the extension of the centrifuge rotor). It was assumed that 7 mm of plasma was formed in the test spin (at $300g$ for 5 min) in 3 mL tubes. It can be seen that the optimal time and length of plasma decrease with increasing effectivity of counter-current, increasing hematocrit and increasing amount of sample, the effect being most pronounced in variation of x_{min}. The extension of the centrifuge rotor at the same length of blood sample had minute effect and the optimal time and the length of plasma slightly increased with increased extension of the rotor (Table 1).

Table 2 presents estimation of parameter ε, optimal time and optimal plasma length in 6 samples of two human donors. The 1st and 2nd generation models were applied corresponding to $\varepsilon = 0$ and ε_{fit}, respectively. ε_{fit} was adjusted for the length of plasma measured in the experiment L_{exp} [obtained after centrifugation at $X = 300g$ for t_{opt} ($\epsilon = 0$)]. ϵ_{fit} was calculated using Eq. (26) and applied to obtain the optimal time $t_{opt}(\varepsilon_{fit})$ with Eq. (24). These results present the test of the 1st generation model and it was found that the length of plasma was overestimated by the 1st generation model. Then, the adjustable parameter ε_{fit} was determined from the experimentally observed length of plasma.

1st generation model pertains to $\varepsilon = 0$ and 2nd generation model pertains to ε_{fit}.

5. Discussion

We have used the 1st generation mathematical model for plasma formation [11] to predict the time and volume (length) of plasma with full recovery of platelets. However, for the estimated optimal time $t_{opt}(\varepsilon = 0)$, the length of plasma L_{exp} observed in 6 samples from 2 donors turned out to be shorter than the theoretically predicted one [$L_{opt}(\varepsilon = 0)$] (Table 2) which points to the inadequacy of the 1st generation model. The discrepancy between the calculated and measured length of plasma was greater

Table 1 Effect of model parameters on estimated optimal time t_{opt} of centrifugation and length of plasma L_{opt}.

Parameter	Ht	X	L_{test} (mm)	x_{min} (mm)	x_{max} (mm)	t_{test} (min)	ε	ω (min^{-1})	t_{opt} (min)	L_{opt} (mm)
ε	0.2	300	7	100	142	5	−0.4	4.51×10^{-5}	13.50	20.04
	0.2	300	7	100	142	5	−0.3	4.51×10^{-5}	13.36	19.81
	0.2	300	7	100	142	5	−0.2	4.51×10^{-5}	13.22	19.59
	0.2	300	7	100	142	5	−0.1	4.51×10^{-5}	13.09	19.37
	0.2	300	7	100	142	5	0	4.51×10^{-5}	12.96	19.16
	0.2	300	7	100	142	5	0.1	4.51×10^{-5}	12.83	18.96
	0.2	300	7	100	142	5	0.2	4.51×10^{-5}	12.70	18.75
	0.2	300	7	100	142	5	0.3	4.51×10^{-5}	12.58	18.56
	0.2	300	7	100	142	5	0.4	4.51×10^{-5}	12.46	18.36
Ht	0.2	300	7	100	142	5	0.2	4.51×10^{-5}	12.70	18.75
	0.25	300	7	100	142	5	0.2	4.51×10^{-5}	12.64	18.66
	0.3	300	7	100	142	5	0.2	4.51×10^{-5}	12.58	18.56
	0.35	300	7	100	142	5	0.2	4.51×10^{-5}	12.52	18.46
	0.4	300	7	100	142	5	0.2	4.51×10^{-5}	12.46	18.36
	0.45	300	7	100	142	5	0.2	4.51×10^{-5}	12.40	18.27
	0.5	300	7	100	142	5	0.2	4.51×10^{-5}	12.34	18.17

(continued)

Table 1 Effect of model parameters on estimated optimal time t_{opt} of centrifugation and length of plasma L_{opt}. *(cont'd)*

Parameter	Ht	X	L_{test} (mm)	x_{min} (mm)	x_{max} (mm)	t_{test} (min)	ε	ω (min^{-1})	t_{opt} (min)	L_{opt} (mm)
x_{min} (mm)	0.4	300	7	80	142	5	0.2	5.59×10^{-5}	16.44	25.41
	0.4	300	7	90	142	5	0.2	4.99×10^{-5}	14.64	22.06
	0.4	300	7	100	142	5	0.2	4.51×10^{-5}	12.46	18.36
	0.4	300	7	110	142	5	0.2	4.11×10^{-5}	9.95	14.37
	0.4	300	7	120	142	5	0.2	3.78×10^{-5}	7.14	10.12
x_{min} (mm)	0.4	300	7	78	120	5	0.2	5.73×10^{-5}	12.05	17.95
	0.4	300	7	83	125	5	0.2	5.40×10^{-5}	12.16	18.06
	0.4	300	7	88	130	5	0.2	5.10×10^{-5}	12.25	18.16
	0.4	300	7	93	135	5	0.2	4.84×10^{-5}	12.34	18.25
	0.4	300	7	98	140	5	0.2	4.60×10^{-5}	12.43	18.33
	0.4	300	7	100	142	5	0.2	4.51×10^{-5}	12.46	18.36

Table 2 Comparison between optimal plasma length predicted by the mathematical model and measured in the experiment.

D	Ht	X	x_{min} (mm)	x_{max} (mm)	ω (min^{-1})	ε_{fit}	t_{test} (min)	$t_{opt}(\varepsilon = 0)$ (min)	$t_{opt}(\varepsilon_{fit})$ (min)	L_{test} (mm)	L_{exp} (mm)	$L_{opt}(\varepsilon = 0)$ (mm)
1	0.42	300	95	142	4.08×10^{-5}	1.77	5	17	12	6	15	21
	0.42	300	96.5	142	4.35×10^{-5}	2.03	5	15	10	6.5	14	20
	0.42	300	100	145	4.82×10^{-5}	2.48	5	13	8	7.5	13	20
2	0.39	300	95	142	9.77×10^{-5}	1.90	5	7	5	15	15	21
	0.39	300	96.5	142	9.03×10^{-5}	0.66	5	7	6	14	18	20
	0.39	300	102	145	8.57×10^{-5}	1.45	5	7	5	14	15	20

D: donor. 1st generation model pertains to $\varepsilon = 0$ and 2nd generation model pertains to ε_{fit}.

in subject 1 in whom the length of the produced plasma in the test sample was shorter (Table 2). It is yet unclear what is the optimal test time or length of plasma, as some discrepancy between the experiment and the theoretical estimation can be expected. It is likely that shorter length is less favorable for testing as it seems that the erythrocyte boundary outside the centrifuge (in the gravity of Earth, $X = 1$) moves in a non-uniform manner as some time is necessary for the plasma to start forming. Once this process begins the boundary moves in a sigmoid-like curve (not shown). These features are not included in the here presented models. In the future, the test procedure should be better defined, however, to elaborate this issue, the existing protocol and models should first be validated on a larger number of samples.

The 2nd generation mathematical model introduced in this work takes into account the effect of hematocrit on the counter-current of plasma at the bottom part of the tube. The velocity of the rising plasma boundary was assumed to be proportional to the velocity of sedimenting erythrocytes. Introduction of hematocrit enabled adjustment of the optimal plasma length, however, to validate the 2nd generation model and suggest further improvements, a larger number of samples with different hematocrit values [14] should be analyzed.

It would be interesting to study whether the effectivity ε is characteristic to a donor and to species. Also the underlying mechanism and the method for estimation of ε should be elaborated. Moreover, the effect of other parameters could be taken into account in the model. Erythrocyte boundary moves towards the bottom of the tube according to a sigmoid-like curve which is approximated by the exponential [15] (Fig. 1C). It is therefore important in which point the test sample is taken.

Processing of blood and plasma may profoundly affect the composition of plasma as cells are prone to shed EVs during the processing due to mechanical, thermal and chemical impacts [16]. Platelets are particularly prone to activate and decompose into EVs [16]. Therefore, other possible impacts [17] should be considered in future higher generation mathematical modeling of particle distribution in therapeutic plasma.

6. Conclusion

By introducing hematocrit into the model, we intended to improve the accuracy of the prediction of optimal centrifugation time and plasma length. We were able to demonstrate that by taking into account hematocrit with adjustable parameter ε we could obtain agreement of experimentally observed and theoretically determined length (volume) of plasma, however at the expense of disparity in estimated optimal time. The model should be tested on a larger cohort of different blood donors including animals to take into account larger interval of different hematocrit values. Autologous plasma is a widely applied therapeutic fluid as its preparation is simple and its use is safe. However, the mechanisms of its effects and of the role of processing procedure are yet obscure. Better understanding of these mechanisms is necessary to optimize the preparation and adjust it to the individual.

Acknowledgments

This research was supported by Slovenian Research Agency J3-3066, P3-0388, J2-4447 and L3-2621, the grant of Nouvelle-Aquitaine for student internship and by Amexio which financially supported the internship of the student.

References

[1] M.R. De Pascale, L. Sommese, A. Casamassimi, C. Napoli, Platelet derivatives in regenerative medicine: an update, Transfus. Med. Rev. 29 (2015) 52–61.

[2] D. Vozel, D. Božič, M. Jeran, Z. Jan, M. Pajnič, L. Pađen, et al., Treatment with platelet- and extracellular vesicle-rich plasma in otorhinolaryngology—a review and future perspectives, Advances in Biomembranes and Lipid Self-assembly, Elsevier, Amsterdam, The Netherlands, 2021, pp. 119–153.

[3] D. Vozel, D. Božič, M. Jeran, Z. Jan, M. Pajnič, L. Pađen, et al., Autologous platelet- and extracellular vesicle-rich plasma is an effective treatment modality for chronic postoperative temporal bone cavity inflammation: randomized controlled clinical trial, Front. Bioeng. Biotechnol. 9 (2021) 677541.

[4] S. Bitenc Zore, D. Vozel, S. Battelino. Facial nerve reconstructive surgery in otorhinolaryngology and its enhancement by platelet- and extracelullar vesicle-rich plasma (PVRP) therapy, in: Proceedings of Socratic Lectures, vol. 7, 2022, pp. 30–37.

[5] J. Etulain, Platelets in wound healing and regenerative medicine, Platelets 29 (2018) 556–568.

[6] P. Everts, K. Onishi, P. Jayaram, J.F. Lana, K. Mautner, Platelet-rich plasma: new performance understandings and therapeutic considerations in 2020, Int. J. Mol. Sci. 21 (2020) 7794.

[7] K. Troha, D. Vozel, M. Arko, A. Bedina Zavec, D. Dolinar, M. Hočevar, et al., Autologous platelet and extracellular vesicle-rich plasma as therapeutic fluid: a review, Int. J. Mol. Sci. 24 (2023) 3420.

[8] M. Yáñez-Mó, P.R.-M. Siljander, Z. Andreu, A.B. Zavec, F.E. Borràs, E.I. Buzas, et al., Biological properties of extracellular vesicles and their physiological functions, J. Extracell. Vesicles 4 (2015) 27066.

[9] S. Fais, L. O'Driscoll, F.E. Borras, E. Buzas, G. Camussi, F. Cappello, et al., Evidence-based clinical use of nanoscale extracellular vesicles in nanomedicine, ACS Nano 10 (2016) 3886–3899.

[10] S.-C. Tao, S.-C. Guo, C.-Q. Zhang, Platelet-derived extracellular vesicles: an emerging therapeutic approach, Int. J. Biol. Sci. 13 (2017) 828–834.

[11] D. Božič, D. Vozel, M. Hočevar, M. Jeran, Z. Jan, M. Pajnič, et al., Enrichment of plasma in platelets and extracellular vesicles by the counterflow to erythrocyte settling, Platelets 33 (2022) 592–602.
[12] N. Steiner, D. Vozel, J. Urbančič, et al., Clinical implementation of platelet- and extracellular vesicle-rich product preparation protocols, Tissue Eng. Part A 28 (2022) 770–780.
[13] N. Steiner, D. Vozel, S. Battelino, Surface-based total blood volume calculation for platelet and extracellular vesicle-rich gel preparation, in: Proceedings of Socratic Lectures, vol. 7, 2022, pp. 39–46.
[14] M. Arko, A. Romolo, V. Šuštar, V. Kralj-Iglič, Role of erythrocyte sedimentation rate (ESR) in preparation of platelet and extracellular vesicles rich plasma, in: Proceedings of Socratic Lectures, vol. 7, 2022, pp. 160–165.
[15] M. Arko, G. Schlosser, A. Iglič, A. Nemec Svete, V. Erjavec, V. Kralj-Iglič, Erythrocyte sedimetation in tubes for preparation of human, equine and canine plasma rich with platelets and extracellular vesicles, in: Proceedings of Socratic Lectures, vol. 8, 2023, pp. 53–57.
[16] V. Šuštar, A. Bedina-Zavec, R. Štukelj, M. Frank, G. Bobojević, R. Janša, et al., Nanoparticles isolated from blood: a reflection of vesiculability of blood cells during the isolation process, Int. J. Nanomed. 6 (2011) 2737–2748.
[17] D. Božič, S. Sitar, I. Junkar, R. Štukelj, M. Pajnič, E. Žagar, et al., Viscosity of plasma as a key factor in assessment of extracellular vesicles by light scattering, Cells 8 (2019) 1046.

Printed in the United States
by Baker & Taylor Publisher Services